MASSIVE MUSCLES

123
456
789
10

IN TEN WEEKS

1

2

3

5

6

8

10

Ed Robinson's muscles are the result of proper training and sensible eating.

MASSIVE MUSCLES IN 10 WEEKS

ELLINGTON DARDEN, Ph.D

DIRECTOR OF RESEARCH
NAUTILUS SPORTS/MEDICAL INDUSTRIES

PHOTOGRAPHY BY CHRIS LUND

A PERIGEE BOOK

Perigee Books are published by
The Putnam Publishing Group,
200 Madison Avenue,
New York, NY 10016

Library of Congress Cataloging-in-Publication Data

Darden, Ellington, 1943 —
    Massive muscles in ten weeks.

    "A Perigee book."
    1. Bodybuilding.   I. Lund, Chris.   II. Title.
[DNLM: 1. Exertion — popular works.   2. Muscles —
growth & development — popular works.   3. Physical
Fitness — popular works. QT 255 D216m]
GV546.5.D36   1986   646.7′5   86-
25403
ISBN 0-399-51340-X

Printed in the United States of America
            9 10

WARNING:
The routines in this book are intended only for healthy
men and women. People with health problems should
not follow the routines without a physician's approval.
Before beginning any exercise or dietary program,
always consult with your doctor.

ACKNOWLEDGMENTS:
Special appreciation is extended to the following
gyms: Gold's Gym of Venice, California; World Gym of
Santa Monica, California; and Gilmore's Gym of De
Land, Florida.

OTHER BOOKS OF INTEREST
by Ellington Darden, Ph.D.:

The Nautilus Bodybuilding Book
    (Revised Edition)
The Nautilus Book
    (Revised Edition)
The Nautilus Woman
    (Revised Edition)
The Nautilus Nutrition Book
The Nautilus Advanced Bodybuilding Book
The Athlete's Guide to Sports Medicine
Strength-Training Principles
Conditioning for Football
High-Intensity Bodybuilding
Super High-Intensity Bodybuilding
The Nautilus Diet

For a free catalog of bodybuilding books, please send a
self-addressed, stamped envelope to Dr. Ellington
Darden, Darden Research Corporation, P.O. Box 1016,
Lake Helen, FL 32744.

CONTENTS

WHAT TO EXPECT

At the start of the program featured in this book, Eddie Mueller stood 5 feet 8 inches and weighed 160 pounds.

Ten weeks later, at the conclusion of the program, he weighed 176 pounds.

Eddie was pleased because he had gained 16 pounds of body weight in just ten weeks. He was also elated because he had added 2 inches to his upper arms, 4 inches to his chest, and 4½ inches to his thighs.

Equally important was the fact that Eddie got leaner. His percentage of body fat decreased from 8 to 6 percent. In the process he lost 2.2 pounds of fat. Thus, Eddie's overall muscles mass gain was 18.2 pounds.

Putting on 18.2 pounds of solid muscle in ten weeks is no easy task. It's very difficult. But it can be done with the right ingredients: *planning, training, eating,* and *resting.*

I personally supervised Eddie

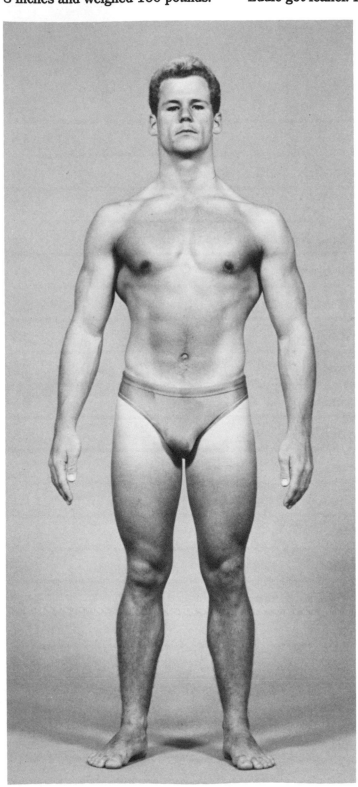

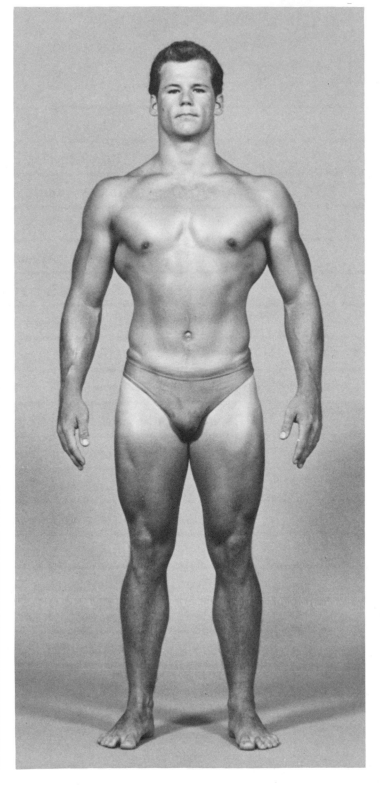

for the duration of the program. I answered his questions. And I kept detailed records of his progress. The results make up a portion of each chapter.

On the following pages you'll find a step-by-step program for getting bigger and stronger in the most efficient manner. Each day for ten weeks you'll be instructed how to exercise and eat for maximum results.

If it's your desire to get massive muscles—and get them in months, not years—then this book is for you.

Thanks to Chris Lund once again for his superb photography, which is used throughout this book. Chris's bodybuilding photos certainly inspired Eddie during his training, and I'm sure they will do the same for you.

Eddie Mueller as he looked before and after the 10-week program featured in this book. Eddie added 2 inches to his upper arms, 4 inches to his chest, and 4½ inches to his thighs.

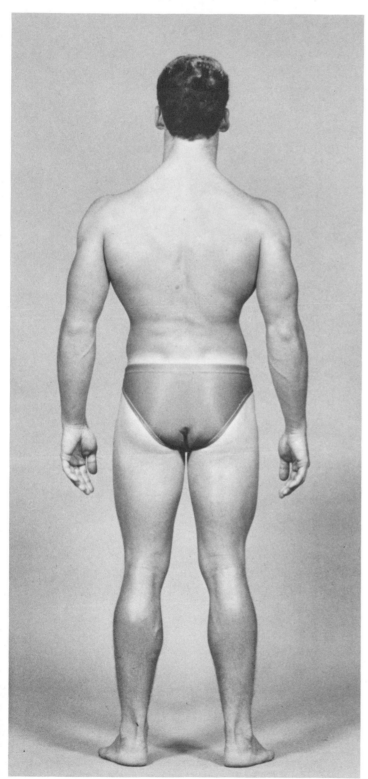

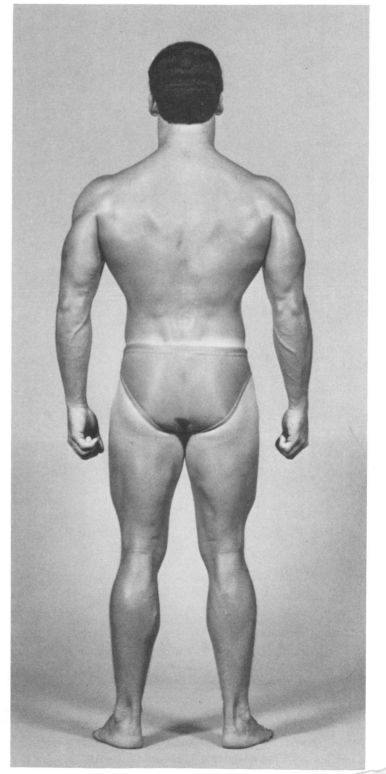

PREPARING FOR MUSCULAR GROWTH

"C'mon, Eddie, one more. Push it up. You can do it."

The young man to whom I was barking instructions was at the brink of exhaustion on the bench press. He was going for his ninth repetition, using a weight that was 50 percent greater than the resistance he could handle nine weeks earlier.

I could see the evidence of our weeks together in the buildup of muscles on his chest. In eight days Chris Lund, a world-famous photographer of bodybuilders, would be in town to photograph Eddie.

With our program near its conclusion, I began to recall where it had started. But first I got Eddie through this day's training session. As soon as he left the gym that evening, I continued to review his progress.

It had begun innocently enough.

"I WANT TO GET BIGGER"

When he walked into my office one afternoon after school, Eddie's greeting was not a hello but a plea. "I want to get bigger," he said firmly.

I looked at him, searching for sincerity. "You and several million other teenagers, Eddie," I said, adding as an afterthought, "How old are you now?"

"Eighteen," he replied.

"How much do you weigh?"

"About a hundred and sixty pounds," Eddie said. "But I'd like to weigh two hundred."

"How long have you been training?"

"Ever since I started high school three years ago," Eddie answered. "And I've gotten bigger—and stronger. I think I've added about fifteen pounds. But it's not enough. I want to be bigger."

"Well," I responded, "what's the

Opposite. The superb physique of professional bodybuilder Ed Kawak. *Above*. Mark Bolduc displays his well developed biceps.

problem? Just train harder and you'll get bigger."

"I can't train the way I want to train," Eddie said. "The coach at school says I should be doing three sets of each exercise and working both my upper body and lower body on the same day.

"Then the guys at my gym say I should be doing split routines, upper body one day and lower body the next. Other people say I should concentrate on powerlifting in order to build bulk.

"Who am I supposed to believe?"

Eddie continued, showing great agony in making his next remark. "I've seen bodybuilders on the beach who tell me I should take steroids. What should I do?"

Before I could answer, Eddie and I were interrupted by a phone call. My secretary informed me that Chris Lund was on the line, calling from England.

"Sorry, Eddie, I have to take this call. But show up here tomorrow and I believe I'll be able to help you."

"Gee, thanks," Eddie said, walking out the door while trying to spread his lats.

"Be here tomorrow at three-thirty," I hollered. Eddie nodded as I turned my attention to the telephone.

A NEW IDEA

"Hello, Chris," I said.

"Hi, Ell," came the reply in a British accent traveling by telecommunication all the way from Tyne and Wear, England. Getting right to business, Chris wanted to know, "Any new ideas for our next bodybuilding book?"

For months, Chris and I had been brainstorming for a third book in our High-Intensity series. Had Chris called ten minutes earlier I might not have had any new ideas. But now one was coming into focus.

"Maybe so, Chris," I responded. "You remember Eddie, that young kid who helped us move Nautilus machines when we photographed Boyer Coe a couple of years ago? Eddie has been after me to put him on a training program. He's got good potential. And, best of all, he seems to be highly motivated."

"Yeah, I remember Eddie," said Chris, who has seen most of the world's greatest bodybuilders through the lens of his camera. "He'd make a good trainee. But how are we going to use him in a book?"

More in a manner of thinking out loud than intentionally speaking, I heard myself tell Chris, "I believe I'll train Eddie on some of our new leverage machines mixed in with basic barbell exercises. I'll organize a ten-week program that will cover his overall body, plus specialized routines for the major muscle groups."

Nautilus leverage machines are plate-loading units that have proven to be very popular in free-weight and bodybuilding gyms. I was beginning to visualize Eddie's program. So was Chris.

"So what you're talking about," Chris replied, "is a ten-week program that applies to beginning and intermediate bodybuilders?"

"Right."

"What about eating?" Chris wondered. "You should cover the important aspects of muscle-building nutrition, especially since a lot of teenagers will be reading the book."

"That's it! That's an excellent idea. That's exactly what we want to do, Chris."

"Bloody good," said Chris. "A step-by-step training and eating program for building massive muscles in ten weeks. It sounds interesting."

"It sure does," I concurred, our enthusiasm becoming contagious. "I'll work out the details with Eddie tomorrow."

A TEN-WEEK PROGRAM

At 3:30 sharp the following afternoon, Eddie was at my door. He was again trying to spread his lats as he walked in.

"Sit down, Eddie. I've got something you should be excited about. I talked to Chris Lund yesterday and we've decided to feature you in our next book. I'm going to put you on a ten-week program."

"Really! That's great," Eddie said enthusiastically. "What's going to be the title of the book?"

"How to Be Skeleton Skinny," I said with a straight face. "Of course, we'll have to get about fifty pounds off your body before we take the final pictures."

Eddie looked puzzled.

"I want to build muscle. I don't want to lose weight, I want to gain it," he said.

"Just kidding, Ed. The book's title will be Massive Muscles in 10 Weeks."

"That's better," said a relieved Eddie. "Will I look like Casey Viator in ten weeks?"

"No, but you'll be a helluva lot more massive than you are now. Maybe we can get you up to a hundred and seventy-five pounds. In six to eight months we could probably get you to two hundred pounds.

"At the very least," I continued, "I guarantee you that when you spread your lats ten weeks from now your back will be wider than your smile."

"Great. When can we get started?"

"Soon, Eddie, soon. But first we need to take a few measurements of your body. I want to have an accurate assessment of your current size and condition."

MEASURING YOUR BODY PARTS

Forget about the 21-inch biceps and the 55-inch chests that the champions claim. Most of these measurements are vast exaggerations. In spite of bogus claims, it is important for you to be aware of your true measurements.

The most important quality of any measurement is its accuracy. You must be precise with your techniques, or you shouldn't bother.

Besides accuracy, you must also be consistent. The specific locations at which you take the measurements must be repeated exactly for the comparisons to be meaningful.

To begin, you'll need a plastic measuring tape that is sixty inches in length. Be sure to label the tape because twenty inches on one tape is not always twenty inches on another. This doesn't matter, of course, if you use the same tape for all your before, after, and in-between measurements.

When taking the measurements, apply the tape lightly to the skin. The tape should be taut but not tight. If you stretch the tape too tight, it will compress the skin and make the value smaller than it actually is. Take duplicate measurements to the nearest one-eighth of an inch at each of the described sites and use the average figure as your circumference score. Do not pump your muscles prior to taking your measurements. Do not take your measurements after a workout. Take all your circumference readings before you exercise.

One final point: it is difficult to take your own measurements. You'll get more accurate values if you have a training partner or friend do them for you.

- **NECK**—Stand in a comfortable position and face forward. Pass the tape around the neck at a level just above the Adam's apple. Make sure the tape is horizontal and the neck is relaxed.
- **UPPER ARMS**—Stand and contract the right biceps. The upper arm should be parallel to the floor. Pass the tape around the largest part of the biceps with the tape perpendicular to the upper arm bone. Measure the left biceps in the same manner.
- **FOREARMS**—Stand and raise your right arm away from your body. Extend your elbow completely, make a fist, bend your wrist, and contract the right forearm muscles. Place the tape around largest part of the forearm, perpendicular to the bones in the lower arm. Measure the left forearm in the same manner.
- **CHEST**—Stand erect. Pass the tape around the back at nipple level and bring it together in the front. Keep the tape in a horizontal plane. Read the measurement at the end of a quiet inspiration of breath and then again at the end of a quiet expiration. The midpoint between the two is the correct measurement.
- **WAIST**—Stand erect and look straight ahead, heels together, with weight distributed equally on both feet. Pass the tape around the waist at navel level. Keep the tape in a horizontal plane. Make the reading at the midpoint of a quiet expiration. Do not pull in the belly.
- **HIPS**—Stand erect and look straight ahead, heels together, with weight distributed equally on both feet. Place the tape around the hips at the position of maximum protrusion of the buttocks. Keep the tape in a horizontal plane and record the measurement.
- **THIGHS**—Stand erect, heels ap-

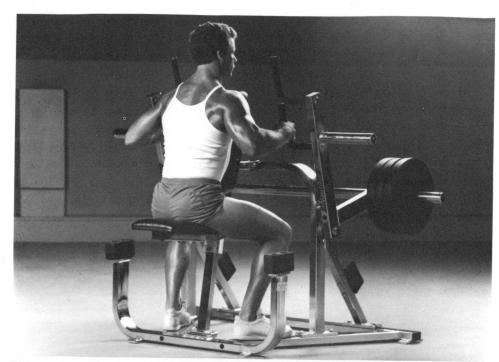

The new Nautilus leverage rowing machine provides meaningful resistance to the biceps and latissimus dorsi muscles.

Erica Giesen, Australian bodybuilding champion, won the 1986 Ms. International title.

Before you start the program, it is important that you accurately record the size of your body.

proximately shoulder-width apart, with weight distributed equally on both feet. Pass the tape around the right thigh just below the buttocks. Keep the tape in a horizontal plane. Do not contract the thigh muscles. Measure the left thigh in the same manner.

• **CALVES**—Stand erect, heels approximately shoulder-width apart, with weight distributed equally on both feet. Pass the tape around the right calf at the widest point. Keep the tape in a horizontal plane. Keep the heels flat on the floor. Do not contract the calf muscles. Measure the left calf in the same manner.

MONITORING YOUR BODY FAT

The easiest and most popular method of calculating the amount of fat you have on your body is by using a skinfold caliper. If you have access to a caliper, use it according to the directions supplied with the device.

Personally, I prefer the Lange caliper. The method I use for men requires me to take the sum of three skinfolds: chest, abdomen, and thigh. For women, I use triceps, hip, and thigh. I then apply this total to a nomogram from the *Research Quarterly for Exercise and Sport* (52:380-384, 1981) to determine the percentage of body fat.

A lean, well-defined bodybuilder will have a body-fat level of under 10 percent. Some champions eventually get below 5 percent.

If you do not have a skinfold caliper available, there's another test you can employ to monitor your body fat. This test will not provide you with a percentage readout, but will let you know whether or not you are getting leaner or fatter. The test involves keeping a periodic record of the difference between your relaxed and contracted upper arm measurements.

In recording your relaxed and contracted arm measurements, apply the following guidelines:
1. Take the measurements before a training session.
2. Relax the arm and take the measurement midway between the elbow and the tip of the shoulder with the arm hanging away from the body. Record the relaxed arm measurement.
3. Flex the arm and measure it at

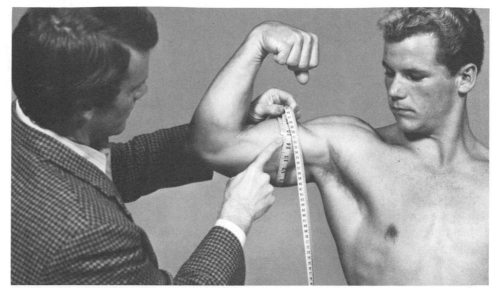
Be sure to keep the tape perpendicular to the bone in the upper arm.

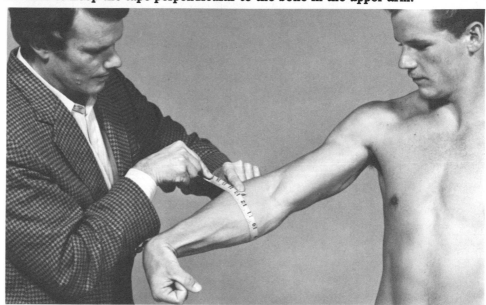
In measuring the forearm, do not bend the elbow. Keep it straight.

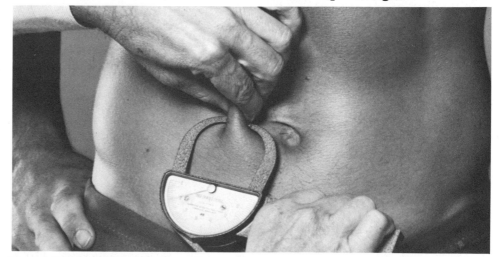
Shown is a skinfold caliper reading being taken on the right side of the navel. The reading, which is 11 millimeters, includes a folded layer of skin and fat. This measurement, combined with others, can be used to determine the amount of fat on the body.

right angles to the bone around the largest part of the contracted biceps with the upper arm parallel to the floor. Record the contracted arm measurement.

4. Take the difference between the relaxed and contracted measurements.

If you are getting leaner, the difference between your relaxed and contracted upper arm measurements will get larger. On the other hand, if you are getting fatter, the difference between the two will get smaller. The reason for the difference is the fact that you *can't flex fat*. Only muscle contains contractile tissue.

Most of your noncontractile fat is stored directly under your skin, with thicker layers around your hips and midsection. When your percentage of fat is reduced, it is reduced to a greater or lesser degree from all over your body.

Thus, by keeping accurate records of the differences between your relaxed and contracted arm measurements, you'll be able to monitor your fatness. If you are getting fatter, then you are consuming too many calories—and you should cut back on your eating.

TAKING FULL-BODY PHOTOGRAPHS

Having a set of full-body photographs made of yourself from the front, side, and back—in both normal and posed states—may be the most meaningful thing you can do to understand your strengths, weaknesses, and overall improvements.

Taking full-length photographs, like those of Eddie Mueller used in this book, is not difficult. But there are important guidelines that must be followed. The most important factor is standardization. Every time you are photographed, you must wear the same bathing suit and employ the same poses, background, lighting, film, camera position, and printing instructions.

1. Have your photographer use a 35-millimeter camera, if possible, and load it with black-and-white or color print film. He should turn the camera sideways for a vertical-format negative.

2. Wear a snug bathing suit or posing suit (a solid color is best) and stand against an uncluttered, light background.

3. Move the photographer away from you until he can see your entire body in the viewfinder. He should sit in a chair and hold the camera level with your navel or, better yet, mount the camera at this level on a tripod.

4. Stand relaxed for front, side, and back pictures. Do not try to suck in your belly.

5. Keep your arms away from your body on the front and back pictures, but against your body on the side pictures. Your heels should be eight inches apart in all three positions.

6. Face the front again and ready yourself for posing. Place your heels eight inches apart. Contract your thighs and then hit a double-armed biceps pose for the camera. Follow this with other poses, such as a front lat spread, front most muscular, side chest, back double biceps, and back lat spread.

7. Have a print made of each negative. On the back of each photograph, write the date, your weight, and any pertinent body part measurement.

8. Repeat the picture-taking session after you've completed the ten-week program. Use the exact same guidelines for all the pictures.

9. Instruct your photography store to make your after prints exactly the same size as your before prints. *Important.* The distance from the top of your head to the bottom of your feet in all the before-and-after photos must be exactly the same for valid comparisons to be made and accurate assessments noted.

MOVING INTO ACTION

Prior to starting the ten-week program for building massive muscles, be sure to take your circumference measurements, body-fat evaluations, and full-body photographs. You'll be glad you did, when—in a few short weeks—you start feeling and seeing your muscles grow.

Another method of monitoring body fat entails recording the difference between the relaxed and contracted upper arm measurements.

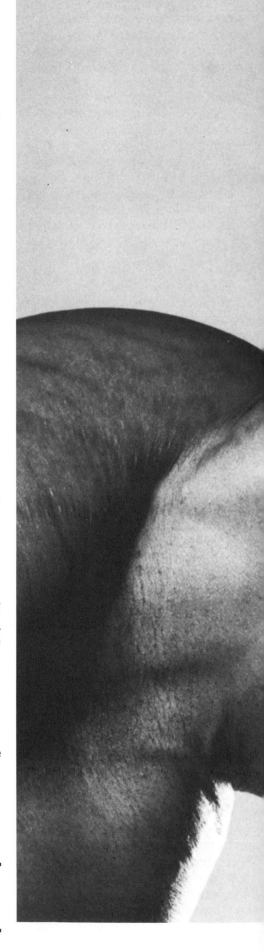

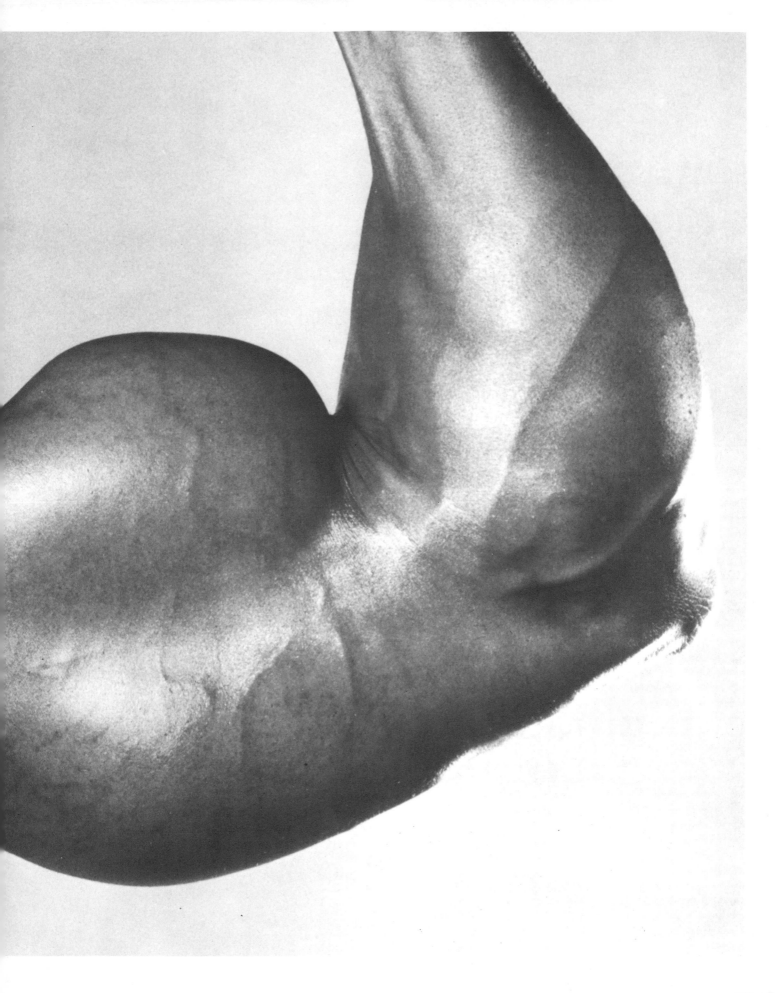

UNDERSTAND-ING TRAINING BASICS

The bottom line is results!

You are training with weights because you want to get bigger and stronger. And you want this size and strength to occur quickly—in months, not years. In minimum time you want maximum results.

Many bodybuilders fail to get maximum results from their exercise because they do *not* apply training basics. Their workouts are *not* based on a solid foundation of facts.

Eddie was no exception. His previous training was riddled with myths and misinformation. Only a few sound physiology principles emerged from his prior exercising.

Perhaps you're in that same boat. If so, now is the time to correct the situation.

INTENSITY

One of the key factors in building massive muscles is the intensity of the exercise. The intensity must be high, involving the ultimate in muscular effort. But exactly what is high-intensity training?

High-intensity training means that you always perform as many repetitions as possible with the appropriately selected weight. Assuming that the resistance has been well chosen, you should barely be able to complete your targeted repetition. But you do not stop at that point. You always try one more repetition. If you are successful, try another. When you are unable to do a complete repetition, stop. You have now reached momentary muscular failure, your goal when performing the exercises in this course.

Above. Lifting and lowering the weight smoothly and slowly are as essential to building massive muscles as the amount of weight you are using. *Opposite.* Form and intensity have always been important factors for Bertil Fox.

In other words, training to momentary muscular failure requires high-intensity muscular contraction, which in turn stimulates a compensatory build-up in the form of added muscle tissue.

Look for ways to make your exercise harder, not easier, and your results will be vastly improved. This is one of the cardinal rules of high-intensity exercise. Of course hard exercise is much less fun than easy exercise, but that's a necessary part of building your body in the most efficient manner. You must simply learn to tolerate the discomfort.

Do not make the mistake of confusing intensity with the amount of exercise. Long, endurance-type exercise cannot be high in intensity. High-intensity exercise, because it is so tiring, must be brief.

When an exercise is performed in a high-intensity manner, one set usually provides your body with optimum stimulation. Generally speaking, multiple sets of the same exercise are not necessary and may even be counter-productive.

PROGRESSION

Another key factor in muscle building is progression. Progression means attempting to increase the workload each training session. You do this by increasing the number of repetitions or the amount of weight—or both.

In the past, I've recommended that bodybuilders usually do eight to twelve repetitions for their upper bodies and fifteen to twenty repetitions for their lower bodies. In 1986, however, after months of testing people with computerized Nautilus machines, we've discovered an individualized way of determining the optimum number of repetitions for each body part. This new concept will be explained fully in the next chapter.

But regardless of the ideal number of repetitions that work best for your

The finely polished physique of three-time Mr. Olympia Frank Zane is a product of progressive resistance exercise.

Bev Francis displays proper form in the dumbbell front raise for her anterior deltoids.

body, one fact remains certain: You still must progress in each of your exercises.

How much should you progress? When you can perform your upper guideline number of repetitions, or more, that is the signal to increase the weight by approximately 5 percent at your next workout.

CORRECT FORM ON EACH REPETITION

Repetitions performed in a smooth, slow manner apply steady force throughout the entire movement. Fast repetitions apply force only at the beginning and the end of the movement. When a barbell is jerked or thrown, a force of three or four times the intended resistance is directed to the joints and muscles. This is both ineffective and dangerous.

The key word in doing a proper repetition with a barbell, dumbbell, or any weight machine is *control*. Control the movement at all times. Do not let the movement control you. You, not momentum, are supposed to be doing the work. If in doubt about your speed of movement, move slower, never faster.

EMPHASIZE THE NEGATIVE

The performance of most exercise requires the raising and lowering of resistance. When you raise the weight, you're moving against the resistance of gravity and performing positive work. Lowering the weight under control brings gravity into play in another fashion. The lowering portion of an exercise is termed *negative work*. During positive work the fibers of the muscles involved in the exercise are shortening. During negative work the same fibers are lengthening.

For building massive muscles, negative work is more important than positive work. Special attention, therefore, should be given to the negative portion of all exercises. A good rule to remember is: Raise the resistance in two seconds; lower the resis-

Bodybuilding champion Juliette Bergman is a believer in emphasizing the negative phase of her exercises.

The shapely backside of Penny Price.

tance in four seconds. In other words, it should take you twice as long to lower a weight as it takes you to raise it.

A few exercises can be performed unassisted in a negative-only manner. For example, negative-only chins and dips can be done by climbing into the top position with your legs and slowly lowering with your arms. Your lower body does the positive work, your upper body the negative work. In negative-only exercise, you can handle 40 percent more resistance than you can use for ten repetitions in a normal positive/negative manner.

CORRECT BREATHING

Beginning bodybuilders are usually concerned about the correct way to breathe while lifting weights. In the past exercise authorities have recommended a variety of breathing techniques during strenuous exercises. Some have suggested breathing in as the weight is lifted and out as it is lowered, while others recommend breathing out as the weight goes up and in as it comes down. With such polarity in recommendations, there is little wonder that novices are confused about how to breathe.

I feel that you should concentrate on your involved muscles during each exercise, not on your breathing. If you simply forget about how to breathe, your breathing will take care of itself and supply your body with adequate oxygen—especially if your heart rate is high enough. There are points along the range of motion of every exercise where it's easier to breathe in or out. Your subconscious mind will identify these points and coordinate your breathing quite efficiently and naturally.

It is essential, however, to refrain from holding your breath during exercise. Keeping your air passages closed while straining can cause something called the "Valsalva effect," which may cause a blackout or a headache. Do not hold your breath. Open your mouth and airway during your exercises and keep breathing.

A good coach or training partner is essential for getting maximum results in minimum time.

DURATION

Fourteen is the maximum number of exercises per workout that you will do during this ten-week course. Approximately one-third of the fourteen exercises are for your lower body, and two-thirds are for your upper body.

A set of ten repetitions should take about one minute to complete. Allowing one minute between exercises, you should be able to perform fourteen exercises in less than thirty minutes. As you work yourself into better condition, the time between exercises should be reduced. Remember, if each exercise is done properly in a high-intensity fashion, brief workouts must be the rule.

WORKOUT FREQUENCY

Rest at least forty-eight hours, but not more than ninety-six hours, between workouts. Exercise, once again, is the stimulation. Your stimulated body must be *permitted* to overcompensate and become stronger. This complex chemical process takes rest, recovery, and time.

An every-other-day, three-times-per-week program also provides your body with the needed irregularity of training. A first workout is performed on Monday, a second on Wednesday, and a third on Friday. On Sunday, your body expects and prepares for a fourth workout, but it does not come. Instead, it comes a day later, Monday, when your body is not expecting it. This schedule prevents your body from falling into a regular routine. Growth is stimulated because your system is never able to adjust to this irregularity of training.

SEQUENCE OF EXERCISE

Most of the time it is important to work your largest muscles first because it causes the greatest degree of overall growth stimulation. However, if you are specializing on a body part, such as your chest, shoulders, or arms—it is appropriate to work the selected muscle group first. Doing so may be just the jolt you need to stimulate new growth.

That's exactly what you'll be doing in the specialized routines in this book. You'll be working your chest first in the chest chapter. And you'll be training your shoulders first in the shoulder chapter. In the overall body chapter, you'll work from larger to smaller, usually in the following order: lower body, torso, arms, waist, and neck.

SUPERVISION

High-intensity exercise is hard work. Few people can do it on their own initiative. You may push yourself to a one-hundred percent effort occasionally, but doing this consistently is virtually impossible.

An instructor or coach could push you through each workout, but this personal supervision is not possible in most circumstances. It is best to choose a friend to be your training partner. Your partner should be dependable and preferably have the same bodybuilding goals as you do.

It's your partner's job to supervise your entire workout. He may frequently tell you to slow down, pause in the contracted position, eliminate excessive back arching, or other activities, each exercise harder and more productive. Your partner will also record the weight and repetitions that you do on each exercise.

After your last exercise, it becomes your responsibility to supervise your training partner's workout. Within several weeks both you and your training partner should be pushing each other through some very productive sessions.

HOW TO CHOOSE A TRAINING PARTNER

A good training partner is worth his weight in gold. You can count on such a person to be dependable, responsible, and most of all, to be available when you need help.

Here's a list of five guidelines that are useful in choosing a training partner:
1.   Your partner should be about your age and strength. The person should have eating habits similar to your own.
2.   Your partner needs to be interested in gaining muscle to the point of making a commitment with you. That commitment needs

to be for at least one hour, three times per week for ten weeks. Your workout, with him supervising, will take less than thirty minutes—and so will his workout, with you doing the supervising. Thus you need to meet at the gym or fitness center at a specific time.
3.   Your partner should *not* be someone you want to impress, but a person you can be honest with.
4.   Your partner should not be the kind of person who judges and criticizes you and offers unconstructive advice.
5.   Your partner should not be your spouse, parent, or other family member. Normal interpersonal problems will get in the way.

Once you have a candidate, talk it over with him. It is important to explain how you want to be treated—that is, strictly or loosely. Then go over your partner's responsibilities. These include showing up for workouts on time, weighing you in, recording your workouts on a card, keeping a daily food diary with you, reinforcing your positive behavior, and understanding your problems.

RECORD KEEPING

Without accurate records of your workouts, it is difficult to measure your progress. That's why it's important to write down your resistance and repetitions after each exercise, or make certain your training partner does this for you. To facilitate the recording of this valuable information, a workout sheet has been placed within each of the ten training weeks. Be sure to use each one.

WARMING UP AND COOLING DOWN

It is always a good idea to warm up prior to heavy exercise as a safeguard against injury. Almost any sequence of light calisthenic movements can be used as a general warm-up to precede your high-intensity workout. Suggested movements include head rotation, hanging from an overhead bar, side bend, trunk twist, and squat. Thirty to sixty seconds of each movement should be sufficient. Specific warming

up of each body part occurs during the first several repetitions of each barbell and machine exercise.

A cool-down period after your workout is also helpful. After your last exercise, cool down by walking around the workout area, getting a drink of water, and moving your arms in slow circles. Continue these easy movements four or five minutes or until your breathing has returned to normal and your heart rate has slowed.

UNDERSTAND AND APPLY

A basic understanding of the training rules in this chapter will lay the groundwork for getting the best possible results from this course. The next chapter covers several new findings concerning repetitions. Study it carefully.

World champion Rich Gaspari is unequaled in his muscular thickness and definition.

MAXIMIZING YOUR IDEAL NUMBER OF REPETITIONS

For more than forty years there has been a controversy among bodybuilders concerning the ideal number of repetitions to perform for building muscles. Some bodybuilders believe that low repetitions, say three to six, are best. Others advocate high repetitions, in the fifteen-to-twenty range. Yet others say that eight to twelve repetitions yield the greatest results.

What is the truth about repetitions? How many should you perform for building maximum muscular size and strength?

The truth is that until January 1986 the answer was purely subjective. No one really knew because there did not

Above. It is important that you determine the ideal number of repetitions to do for your arms, torso, and lower body. **Opposite.** Your biceps may require a different repetition scheme for growth stimulation from that for your lats and quadriceps. To find out, you must test your degree of *inroad* in certain exercises.

exist a valid way to measure the physiological effect on your body from different repetition schemes.

Now there is a way to measure accurately the effect each repetition of an exercise has on the involved muscle.

A NEW DISCOVERY

In January of 1986, Arthur Jones, the inventor of Nautilus, introduced a new testing and training machine to the exercise world. The device, officially called the Leg Extension Medical Machine, consists of a very sophisticated leg extension for the quadriceps. It has a computer tie-in that measures force output and position of the leg. The machine computes a strength curve so that you can see not just your peak strength, but also your strength at every point throughout the entire range of motion.

Before I describe the testing procedures used with the new medical machine, let me first tell you about the significance of the testing and how this concerns building muscle.

THE IMPORTANCE OF INROAD

Arthur Jones now feels that a key factor in muscular growth stimulation is *inroad*. Inroad is the depletion of momentary strength from a set of an exercise. He believes that the proper inroad that stimulates the fastest muscular growth is approximately 20 percent for all major muscle groups. In other words, the number of repetitions that you perform should reduce your starting level of strength by 20 percent.

Jones first became aware of the relationship between positive strength and endurance more than twenty years ago. The following excerpt is taken from his article, "Exercise . . . 1986," and deals with his experience with barbell exercises for his upper body:

> When I was able, but barely able, to lift 100 pounds only once, then I knew that I could perform exactly ten repetitions with 83 pounds . . . no more, no less, not nine, not eleven, exactly ten.
>
> When I was able to perform ten repetitions with 100 pounds, then I knew that I could perform one rep-

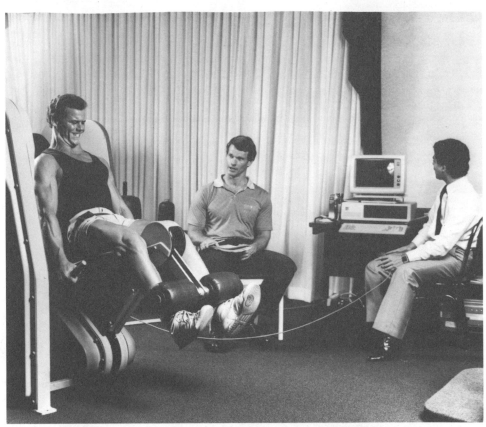

Eddie Mueller is being tested on the Leg Extension Medical Machine. The machine measures the strength of Eddie's quadriceps in seven different positions and then provides him with a computer printout of his strength curve.

etition with exactly 120.

When my strength changed, up or down, my anaerobic endurance went up or down to exactly the same degree.

Thus, when I reached a point of momentary muscular failure, after having performed exactly ten repetitions with 83 percent of my starting level of positive strength . . . then, at that point in the exercise, I failed because my remaining strength was slightly below 83 percent of my starting level.

So ten repetitions with 83 percent of my starting strength level reduced my strength, momentarily, by about 18 percent.

Being clearly aware of this ratio in myself, I assumed that it applied to other people as well . . . thus, when I encountered a man who could perform only four repetitions with 80 percent of his starting level, I at first accused him of not trying hard enough.

But he was trying, although I failed to recognize this at the time; that was all he could do; his ratio

was different.

Later, another man performed twenty-three repetitions with 80 percent of his starting strength level, and again I was surprised.

These and other examples finally made me aware of the fact that this ratio varies on an individual basis . . . but I simply overlooked what should have been the obvious implications.

For years, many more years than I even like to remember, I have been telling people to train in much the same way; select a weight, by trial and error, that will permit you to perform seven or eight repetitions in good form; but continue for as many repetitions as you can in good form, stopping only when it becomes momentarily impossible to continue.

Then, during later workouts, always perform as many repetitions as possible . . . but when it becomes possible to perform ten or more repetitions in good form, then increase the weight about 5 percent.

Never use a weight that will not

permit at least seven repetitions, and always increase the weight slightly when it becomes possible to perform ten or more repetitions.

Most other people have been giving very similar advice for the last fifty years . . . and millions of people have trained in this manner.

But I now understand that this advice is wrong . . . wrong for many people; and I understand why it is wrong.

MY EXPERIENCE WITH INROAD

I understand what Jones is saying. And I've been guilty of recommending the same basic repetition scheme, eight to twelve, for everyone. But my research and experience have shown me that approximately 70 percent of the bodybuilders who train with eight to twelve repetitions get good results. The remaining 30 percent get little or poor results, which isn't surprising when I analyze the situation.

The whole idea behind doing a given number of repetitions of an exercise is to make the correct inroad—enough of an inroad to stimulate growth, but not so much that you prevent it from occurring.

INROAD COMPARISONS

One testing and research leg extension unit, for example, can be programmed to display the strength curve on the computer screen while the subject is performing the positive phase (lifting) of the exercise. As the subject does each repetition, the foot-pounds of work performed throughout the full range of movement are calculated. Thus, by looking at the computer screen, you can observe what happens to the trainee's strength repetition by repetition. Furthermore, if you consider the first repetition to be 100 percent of the subject's strength, then each succeeding repetition can be evaluated as a percentage of total strength. This indicates the inroad, the depletion of strength caused by succeeding repetitions.

Please examine the following two charts. They were derived from computer printouts of two different subjects as they exerted maximum effort

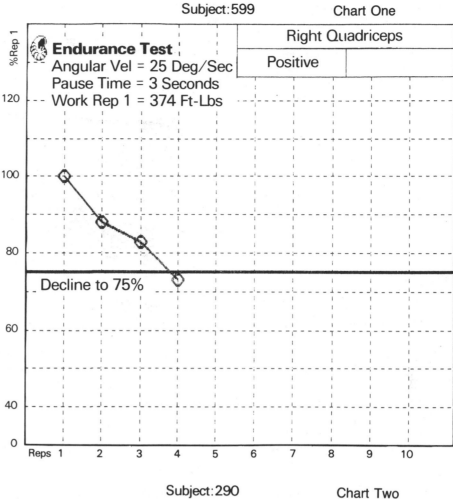

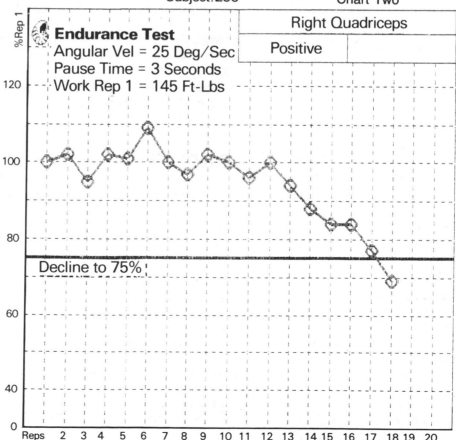

on every repetition of the leg extension testing and research unit.

Chart 1 shows a subject who could perform only three maximum repetitions before he fell below 75 percent of his first repetition. Each repetition made an average inroad of 9 percent into this subject's starting level of strength.

In contrast to the subject in Chart 1, the trainee in Chart 2 required seventeen repetitions to fall below the 75 percent level. Each repetition made a 1.75 percent inroad into this subject's starting level of strength.

In other words, both trainees made approximately the same inroad, 27 percent and 30 percent respectively, while doing the same exercise. But it took the second subject 5.67 times as many repetitions as the first subject.

NEUROLOGICAL EFFICIENCY

Arthur Jones calls this phenomenon "neurological efficiency," or the hook-up between the brain and muscles. All reports indicate that neurological efficiency is genetic and not subject to modification through training. But it is evident that there is great variation among people in their neurological efficiency.

A bodybuilder who has a high level of neurological efficiency will be able to contract a large amount of his muscle in an all-out effort. He will be much stronger than average, but his muscular endurance will be lower than average. His leg extension endurance test would resemble Chart 1.

On the other hand, a bodybuilder with low neurological efficiency will be weaker than average. His muscular endurance, however, will be greater than average. Chart 2 would be typical of his endurance test.

THE 20 PERCENT GOAL

The main point behind the discussion in this chapter so far is that the repetition plan that is correct for one bodybuilder may be totally wrong for another. High repetitions are an absolute requirement for growth stimulation for the trainee in Chart 2. But high repetitions for the subject in Chart 1 would constitute gross overtraining, and the inroad would be

much greater than the 20 percent goal.

The low repetitions that work successfully for the Chart 1 trainee would have no effect on the subject in Chart 2. Four or five repetitions would make no meaningful inroad into the second subject's strength level, and thus it would not provide growth stimulation.

(As an interesting note, the individuals in Chart 1 and Chart 2 are not at the far extremes of the several hundred people whom we have tested for muscular endurance. At the extreme left of the range, we have encountered one subject who could perform only one repetition with 80 percent of his one-time maximum. At the extreme right of the range, we tested a subject who performed thirty-four repetitions with more than 80 percent of his maximum. The average number of repetitions of most people tested on the leg extension is eight, with the typical span running from six to eleven. Those doing fewer than six repetitions appear to have high neurological efficiency, while those performing more than eleven repetitions seem to have low neurological efficiency.)

So the goal for maximum growth stimulation, once again, is to make a 20 percent inroad into your starting level of strength. But if you do not have access to the new Nautilus Medical Machines, how do you know the proper number of repetitions to do on the various exercises that make up your routine?

TESTING YOURSELF WITH STANDARD EQUIPMENT

Don't fret! There are several tests that you can perform on the equipment that you presently use. To determine your guideline range of repetitions to employ on any barbell or weight-machine exercise, follow these steps:
1. Determine your one-repetition maximum on any exercise.
2. Rest at least five minutes.
3. Take 80 percent of this one-repetition maximum and perform as many repetitions as possible in proper form. Do not cheat.
4. Make a written note of the number of repetitions.

5. Multiply the number of repetitions by .15.
6. Round off the resulting figure to the nearest whole number.
7. Add this whole number to your 80 percent repetitions. This becomes the high end of your repetition guideline.
8. Subtract the same number from your 80 percent repetitions. This becomes the low end of your repetition guideline.

TESTING EDDIE MUELLER

Using the above steps, I tested Eddie Mueller at the start of his ten-week program. We used eight Nautilus leverage machines and three barbell movements. The chart below details the results and subsequent repetition guidelines.

REPETITION GUIDELINES FOR EDDIE MUELLER FOR ELEVEN EXERCISES (Before Ten-Week Program)

Exercise	One-Rep Maximum	Rep with 80% of Maximum	Rep Guideline (±.15 of 80% Maximum Rep)
Nautilus Leverage Machine			
Leg extension	170/1	135/7	6–8
Leg curl	155/1	125/7	6–8
Leg press	230/1	185/9	8–10
Pullover	150/1	120/6	5–7
Bench press	200/1	160/7	6–8
Rowing	240/1	190/6	5–7
Biceps	85/1	70/6	5–7
Triceps	80/1	65/6	5–7
Barbell			
Squat	250/1	200/11	9–13
Standing press	155/1	125/7	6–8
Standing curl	135/1	110/7	6–8

Let's take the leg curl as an example of how the chart is designed. Eddie's one-repetition maximum is 155 pounds. Thus, we give him 80 percent of 155, or 124, which is rounded off to 125 pounds. He then does as many movements as possible in good form —which is 7 repetitions. Now we multiply .15 by 7, which is 1.05, which rounded off becomes 1. To 7 we add 1 and to 7 we subtract 1, with the resulting range being 6 to 8.

Eddie's guideline range of repeti-

Past reports of Bev Francis's strength indicate that she has high neurological efficiency in her upper body.

tions, or the repetitions he must use to make approximately a 20 percent inroad into his hamstring strength, is 6 to 8. Anytime he can perform 8 or more repetitions, that is the signal to increase the leg curl weight at his next workout by 5 percent.

MULTIPLE-JOINT EXERCISES TEST HIGHER

You'll notice that Eddie's performance with 80 percent of his maximum, on seven of the eight leverage machines, is either six or seven repetitions. On the leg press, which is a multiple-joint movement, he did nine repetitions. It is not unusual for a trainee to require higher repetitions on the leg press because it is a multiple-joint movement that involves a rest period each time the knees are locked out.

It is also interesting to note that Eddie did a repetition or two more on the barbell exercises than he performed on similar leverage machine movements. Barbell exercises, because of their inherent limitations, do not involve as many muscle fibers as similar exercises performed on leverage machines. Thus it was expected that Eddie would do a few more repetitions on comparable barbell exercises, and he did.

THREE IMPORTANT BODY PARTS

The recommended testing procedure can be applied to determine your ideal repetition plan on almost any barbell, dumbbell, or weight machine exercise. Just remember to keep your form very strict.

I have found, however, that you do not need to perform the test with all possible exercises. The muscles of

Left. "My triceps need some attention," Danny Padilla seems to be saying to Frank Richard. The best way to give the triceps attention is first to determine the ideal number of repetitions to do for maximum stimulation.

Opposite. Most trainees require higher repetitions for multiple-joint exercises, such as the leg press and the squat, than they do for single-joint exercises, such as the leg extension and the leg curl.

your lower body seem to be very similar in their repetition needs, as are your torso muscles, and your arm muscles. By selecting one or two exercises from each of the three areas—lower body, torso, and arms—you should be able to test and calculate repetition guidelines that can then be applied to all your other exercises.

Most of Eddie's exercises are performed in the range of five to eight repetitions. He does slightly higher repetitions on the leg press and squat, as well as the calf raise. Other bodybuilders I have worked with, however, require different repetition patterns for the three main areas. For example, one bodybuilder I recently tested needed fourteen to eighteen repetitions for his lower body, nine to thirteen repetitions for his torso, and five to seven repetitions for his arms. His training partner required eleven to fifteen repetitions for his lower body, eight to ten repetitions for his torso, and nine to thirteen repetitions for his arms. Few bodybuilders test the same throughout their muscle groups. There are usually slight variations, with higher repetitions required for the low-

er body. But there are many exceptions.

The only way that you can be certain about the best repetition guidelines to employ for maximum results is to test yourself using the previously described steps.

You'll be glad you did after you see the results that the testing and subsequent repetition guidelines will make on your quest for massive muscles.

Above. The Gold Gym's A Team, as they appeared during Chris Lund's recent trip to California. It is possible that each man requires different repetitions for maximum growth of his biceps.

Opposite. Frank Richard of England made his mark on the professional bodybuilding circuit in 1986.

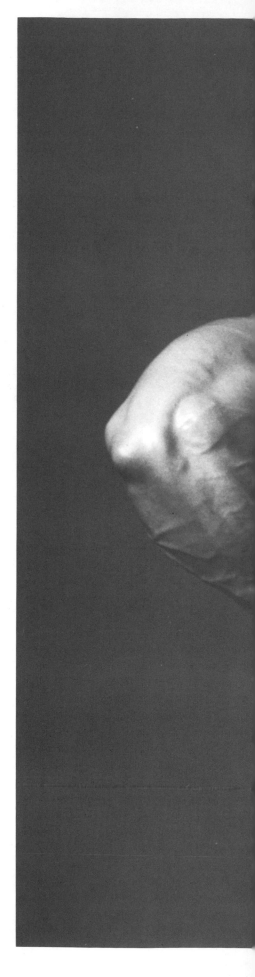

APPLYING MUSCLE-BUILDING NUTRITION

Building massive muscles involves three factors: exercise, rest, and food. Too much or too little of any of the three factors will limit or prevent growth. Maximum muscular growth requires the right amount of all three.

The guidelines for exercise and rest were discussed in the last two chapters. Now we'll turn our attention to the foods that are necessary for building muscle.

The science of nutrition—the study of food and how the body uses it—will not be covered comprehensively in this book. I've already done that in *The Nautilus Nutrition Book*, so I'll simply refer you to it. But I would like to say that what you read in muscle magazines about the need for protein and amino acid supplements, vitamin and mineral pills, and certain "health" foods is seldom based on scientific studies. Many of the writings are com-

posed of half-truths, misinformation, and outright fraud, and they are cleverly designed to promote mail-order food supplements sold through the magazines.

To build muscle, you do not need to spend your money on expensive food supplements. In fact, quite the contrary is true. You can build massive muscles from common foods that you purchase at reasonable prices from your local supermarkets.

THE SIGNIFICANCE OF CALORIES

The key food-related component in

Above. A muscle-building diet includes several servings a day from the Basic Four Food Groups: meat, milk, fruit/vegetable, and bread/cereal. *Opposite.* The mass and symmetry of John Terilli are an indication that he understands that *calories count*!

Nicole Bass shows excellent muscularity in her back.

muscle is calories. There are 600 calories in a pound of muscle. This may seem rather low until you realize that 70 percent of a muscle is made up of water, and water contains no calories.

Adding calories to your diet will not make your muscles larger, however, unless you've stimulated them to grow at the basic cellular level with proper exercise. Stimulation must always precede calories.

Calories are a measure of food energy. The primary energy-supplying foods are carbohydrates, fats, and proteins.

Fats contain more than twice as many calories as an equal amount of protein and carbohydrates.

1 gram of fat	= 9 calories
1 gram of protein	= 4 calories
1 gram of carbohydrate	= 4 calories

Your body, however, does not prefer fats as a source of calories for building muscle. And contrary to popular bodybuilding lore, it does not prefer proteins either. It prefers carbohydrates. Carbohydrates are most efficiently converted by your body to the raw materials that are necessary to build muscle. Naturally, your body also needs a certain amount of proteins, fats, vitamins, minerals, and water to make the process complete. But most of all it needs calories, calories from carbohydrates.

EDDIE'S CALORIES

In preparing Eddie's body for muscular growth, I made sure he was aware of the number of calories that he had been consuming each day. Was he eating too much, too little, or just right?

Also, what about the composition of his meals? Was he getting enough carbohydrates and not going overboard on proteins and fats?

To determine the answers I asked Eddie to keep an accurate record of everything he ate and drank (including water) for three days. When he had done this, I sat down with him and analyzed his intake using the *USDA Handbook No. 456: Nutritive Value of American Foods.* Briefly, here was what we found:

Eddie consumed an average of 3,038 calories per day. This was about right for him to maintain his body weight of 160 pounds. To estimate the number of calories that a young man of Eddie's age needs to maintain his weight, most nutritionists recommend that you multiply your body weight in pounds by a number from 15 to 20. Use 15 if your typical activities during the day are light and mostly sedentary. Use 20 if you are very active, and 17 or 18 if your level of activity is moderate.

We used 19 for Eddie (19 × 160 = 3,040).

Eddie's calorie composition was 39 percent fat, 16 percent proteins, and 45 percent carbohydrates. Again, this was fairly typical of a teenager's diet.

Clarence Bass, known by many people as "Mr. Ripped," nourishes his body with a breakfast of cereal and fruit.

Such a diet, in my opinion, contains too many fats and proteins, and not enough carbohydrates. Primarily, Eddie needed to eat more fruits, vegetables, breads, and cereals—and fewer meats and dairy products. I wanted Eddie to consume a diet each day that was composed of approximately 28 percent fats, 13 percent proteins, and 59 percent carbohydrates. Accomplishing such an ideal would be most conducive to building massive muscles.

Eddie desired to gain body weight by building his muscles. He did not want to get fat. If anything, he wanted to get leaner—to get more definition.

Eddie's goal was to gain from one to two pounds of muscle per week for ten weeks.

PROGRESSIVE PLANNING

With the help of a meal-planning specialist, we worked up some basic menus for Eddie to follow. He would eat three meals a day, plus consume a nutritious milkshake snack.

Most important, each week as the program progressed, Eddie added 100 calories to his basic diet. Here's how the progressions looked in chart form.

Week 1: 3,100 calories per day
Week 2: 3,200 calories per day
Week 3: 3,300 calories per day
Week 4: 3,400 calories per day
Week 5: 3,500 calories per day
Week 6: 3,600 calories per day
Week 7: 3,700 calories per day
Week 8: 3,800 calories per day
Week 9: 3,900 calories per day
Week 10: 4,000 calories per day

The ascending-calorie diet worked well for Eddie. And it will work for you, too. Depending on your age, height, weight, and activity level, you may have to juggle the daily calories up or down. But it shouldn't vary dramatically from Eddie's regime.

The main concepts that you should understand are as follows:
1. Determine the number of calories that you need daily to maintain your body weight. This is estimated by multiplying your body weight in pounds by 15 to 20.
2. Add 100 calories per day to this total for the first week. Then add 200 calories per day for the second week, and 300 calories for the third, and so on for ten weeks.
3. Keep your daily calories composed of an approximate 59 percent carbohydrates, 28 percent fats, and 13 percent proteins.

All of this has been done for you, in both a general and specific fashion, in the week-by-week chapters. You simply follow the guidelines and sample menus and make a few substitutions and adjustments when necessary.

BASIC FOUR FOOD GROUPS

The sample menus presented in the ten-week program are designed using the Basic Four Food Groups and a 4:4:8:8 ratio. The 4:4:8:8 ratio is a quick-check reminder of how much of each food group you should have as a minimum. For example, the basic eating plan for one day translates to:

- 4 servings from the Meat/Poultry/ Fish/Legumes (dried beans, peas, lentils)/Eggs Group—referred to in abbreviated form as the Meat Group.
- 4 servings from the Milk/Yogurt/ Cheese Group—referred to as the Milk Group
- 8 servings from the Fruit/Vegetable Group
- 8 servings from the Bread/Cereal Group (which includes rice and pasta)

Sticking to the 4:4:8:8 ratio, along with certain "other foods," assures that your daily calories are composed of approximately 50 to 60 percent carbohydrates, 25 to 30 percent fats, and 12 to 15 percent proteins. Such a composition, once again, is most conducive for building massive muscles.

BASIC FOUR, PLUS "OTHER FOODS"

The newest version of the Basic Four Food Groups also pays nutritional heed to a fifth food group: fats, sugars, and alcohol, sometimes (like the chart above) referred to as "other foods." These influence your diet categorically as well as nutritionally. They also add taste and satisfaction to your meals.

When watching your waistline, these are the items best reduced without fear of loss of important nutrients. But since these foods add flavor and fun, there is no need to eliminate them from your muscle-building diet. Simply consume them in moderation.

PUTTING THE PLAN INTO PRACTICE

You are now ready to progress to Week 1 of the ten-week program. At first glance the routine may appear to be too simple. Do not let simplicity fool you.

With the right planning, testing, and pushing, that simple routine will soon become brutally hard, precisely stimulating, and extremely productive.

Let's get started!

To keep your waistline lean, you must limit your consumption of foods that are high in calories but low in nutrients.

BASIC FOUR FOOD GROUPS FOR BODYBUILDERS

BASIC FOOD Group	MINIMUM DAILY Servings	Serving Size	Food Sources
Meat	4	2–3 ounces cooked 2 medium 2 tablespoons ½ cup 1 cup	Meat, poultry, fish Eggs Peanut butter Cottage cheese Dried beans or peas
Milk	4	1 cup 1½ ounces 1–1¾ cups	Milk, yogurt Cheese Milk-containing foods
Fruit/Vegetable	8	½ cup raw or cooked ½ cup juice 1 cup raw	Fruit or vegetable Fruit or vegetable Dark green leafy or yellow vegetable
Bread/Cereal	8	1 slice ½–¾ cup ½ cup ½ cup	Breads: whole-grain and enriched, muffins, rolls Cereals: cooked, dry, whole-grain, grits, barley, flours Pasta: macaroni, noodles, spaghetti Rice: brown or white
Other Foods	6*	1 teaspoon 1 teaspoon 2 teaspoons	Butter, margarine, oil Salad dressing Jellies, jams, and other sweet toppings Alcohol (not recommended)

*May be adjusted up or down depending on daily caloric needs.

WEEK 1: ESTABLISHING YOUR BASIC ROUTINE

I was outlining on paper the opening week of Eddie's training, and in his mind Eddie was outlining what he wanted to accomplish. He had seen a glimpse of the physique he wanted to achieve for himself.

"I'll never forget the first time I saw Casey Viator," Eddie said, carrying a reverence in his voice worthy of the youngest Mr. America in history, an accomplishment of Viator's in 1971.

Eddie was an elementary school student in Lake Helen when Viator was in training at Nautilus headquarters.

"It was during recess one morning," recalled Eddie. "We were playing out by the front gate. Along came the big-

Above. The barbell squat is one of the very best basic exercises for building massive muscles. *Opposite.* This is how Casey Viator looked in 1978 when Eddie Mueller saw him ride a bicycle in front of the Lake Helen Elementary School.

gest arms and legs I had ever seen on a man, and he's riding a bicycle. I just stopped and stared as he rode by. We all did!"

Eddie's recollection jogged my memory.

"That must have been 1978," I said. "You were probably in the fourth or fifth grade. Casey was training for the Mr. Universe contest that year and he was trying to burn a few extra calories by riding his ten-speed bicycle twice a day."

Eddie contracted his arms with a dream bursting from his heart. "Boy, Casey was sure massive. From that day on I wanted to look like him."

I nodded my agreement concerning Casey's physique, and then seized the opportunity to make a valuable point with Eddie.

"You know, Eddie, one of the primary reasons behind Casey's muscular body was the fact that he had a basic mass-building routine that he kept going back to—month after month, year after year. Sure, he specialized occasionally on certain body parts, but much of his training consisted of plain vanilla routines. And he always trained with heavy poundages.

"What I want to do for the first two weeks of this program," I continued, "is to put you on a basic workout—a basic mass-building routine that you can keep coming back to for the rest of your training days."

Eddie was eager. "Okay," he said, "let's tackle the basics."

Prior to discussing the basic routine, let's detail the format for Weeks 1–10. Each chapter will start with a brief introduction. Following the introduction will be a general listing of servings from the Basic Four Food Groups to eat each day. Under the listing will be a sample menu for one day with all the calories noted. At the bottom of the column, there will be a workout card for your use. Finally, the specific training routine will be discussed and illustrated.

The impressive back of Casey Viator.

GENERAL SERVINGS AND SAMPLE MENU FOR WEEK 1

GENERAL SERVINGS (TOTAL CALORIES: 3,100 PER DAY)

Basic Food Group	Recommended Daily Servings
Meat	5.00
Milk	5.00
Fruit/Vegetable	10.00
Bread/Cereal	10.00
Other Foods	7.50

SAMPLE MENU (TOTAL CALORIES: 3,100 PER DAY)

Breakfast	Calories:	790
3 slices of bacon, crisp and drained		135
2 eggs, any style		160
3 slices of bread, toasted		240
4 teaspoons of jelly or preserves		70
1½ cups of milk, skim		135
½ cantaloupe		50

Lunch	Calories:	970
2 sandwiches		
1 peanut butter: 2 slices of bread		160
2 tablespoons of peanut butter		180
1 grilled cheese: 2 slices of bread		160
1 teaspoon of butter or margarine		35
1½ ounces of cheese, swiss or cheddar		150
1 cup of orange juice, unsweetened		90
1 pear		100
½ cup of corn, cream-style		95

Dinner	Calories:	1035
6 ounces of fish, broiled with lemon juice		160
1 baked potato		150
3 tablespoons of sour cream		100
3 slices of bread		240
2½ teaspoons of butter or margarine		90
1 salad: 2 cups of lettuce, torn		20
1 tomato, sliced		35
2½ ounces of dates		240

Snack	Calories:	305
Strawberry milkshake: 1 cup of milk, whole		150
½ cup of ice milk, strawberry		100
1 cup of strawberries, fresh or frozen		55

WEEK 1: BASIC ROUTINE

Date:			
1. Leg extension			
2. Leg curl			
3. Full squat or leg press			
4. Reverse leg raise			
5. Calf raise			
6. Lateral raise			
7. Overhead press			
8. Nautilus pullover			
9. Pulldown behind neck			
10. Bench press			
11. Bent-over row			
12. Triceps extension			
13. Biceps curl			
14. Stiff-legged deadlift			

Above. The leverage lateral raise machine places the resistance directly on the upper arms.
Below. Nautilus pullover: Exert force with your elbows, not your hands, and you'll feel your lats working throughout the full range of motion.

TRAINING GUIDELINES FOR BUILDING MASSIVE MUSCLES

1. Perform no more than a total of fourteen sets of all exercises in any one training session. Follow the routines exactly as listed, week by week, for ten weeks.

2. Train no more than three times a week. Each workout should involve your entire body, as opposed to splitting the routine into lower- and upper-body work on separate days.

3. Select a weight, or level of resistance, for each exercise that allows you to fall between your tested low and high repetitions guidelines.

4. Increase the resistance by approximately 5 percent at your next workout, when more than the high guide number of repetitions are performed.

5. Accentuate the negative or lowering portion of each repetition. Lift the weight in two seconds and lower it in four seconds.

6. Move slower, never faster, if in doubt about the speed of movement.

7. Attempt constantly to increase your number of repetitions or the amount of weight, or both. But do not sacrifice form in an attempt to increase your repetitions or your weight.

8. Get ample rest after each training session. High-intensity exercise necessitates a recovery period of at least 48 hours. Your muscles grow during rest, not during exercise.

9. Train with a partner who can reinforce proper form on each repetition.

10. Keep accurate records—date, resistance, and repetitions—of each workout.

BASIC MASS-BUILDING ROUTINE

Before beginning the recommended routine, make sure you've accomplished these tasks:

- Body-part measurements
- Skinfold readings or difference between relaxed and contracted arm measurements
- Full-body photographs
- Repetitions guideline tests for your lower body, torso, and arms

Without proper evaluations, you'll be training without a good support

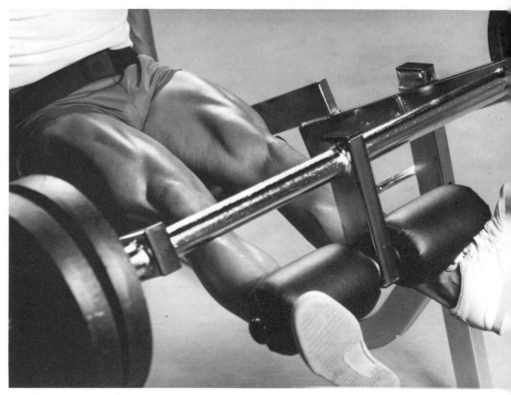

Above. **Leg curl: Do not bounce in and out of the contracted position. Make your hamstrings do all the work.**
Below. **Leg extension: Stop completely in the top position and your quadriceps will be involved to a higher degree.**

system. Get your measurements and testing done now.

Your basic mass-building routine is as follows:

1. Leg extension
2. Leg curl
3. Full squat or leg press
4. Reverse leg raise
5. Calf raise
6. Lateral raise with dumbbells
7. Overhead press
8. Nautilus pullover
9. Pulldown behind neck
10. Bench press
11. Bent-over row
12. Triceps extension with one dumbbell held in both hands
13. Biceps curl
14. Stiff-legged deadlift

● LEG EXTENSION: This is the best exercise for your quadriceps or frontal thigh muscles. Sit in the leg extension machine and place your feet behind the roller pads. If possible, align the axis of rotation of the machine with your knees. Lean back and stabilize your body by grasping the sides of the machine. Straighten your legs smoothly. Pause at the top, in the position of full muscular contraction. Do not bounce in and out of the top. Lower slowly and repeat.

● LEG CURL: This is the most productive movement for your hamstrings. Lie face down on the leg curl machine and place your heels under the roller pads. Make certain that your knees are in line with the axis of rotation of the movement arm. Bend your legs, trying to touch your heels to your buttocks. In the fully contracted position, your hips should be raised, and you should come to a complete stop. Lower slowly and repeat.

● FULL SQUAT OR LEG PRESS: Choose either the full squat or leg press. Do not do both on the same training day.

Full squat: The barbell should be in the top position of the squat racks. Step under the barbell and place it behind your neck and across your shoulders. Stand erect and step back. Your

---

Always perform each repetition in a steady manner. Never begin with a jerk. Look for ways to make your repetitions harder, not easier.

You may substitute the bent-arm pullover with a barbell for the Nautilus pullover. From the stretched position, pull the barbell over your face to your chest. Lower to the floor and repeat.

The leverage bench press machine is used in a manner similar to the bench press with a barbell. The leverage machine, however, is much safer to use since the weight cannot be dropped across the chest.

feet should be approximately shoulder-width apart, and your head up at all times. Bend your knees and hips, and lower your buttocks slowly until the back of your thighs touch your calves. Do not bounce in and out of the bottom. Return to the standing position in a smooth fashion. Take a deep breath and repeat.

Leg press: Adjust the seat in the leg press machine to where you have the greatest range of movement. Most bodybuilders move the seat back, rather than pulling it forward. Sit in the leg press machine and place both feet on the movement platform. Your feet should be about shoulder-width apart. Straighten your legs smoothly. Slowly lower and repeat.

● REVERSE LEG RAISE: Performed immediately after squats or leg presses, the reverse leg raise really works the contracted position of your buttocks and lower back muscles. Lie face down on the floor. Anchor your upper body by holding on to a sturdy support. Lift both legs backward off the floor as high as possible. Keep your knees and heels together. Pause in the highest position and squeeze your buttocks together. Lower your legs slowly to the floor and repeat.

● CALF RAISE: You'll need a calf raise machine of some sort for this movement, or you may perform the exercise with a barbell that has been incorporated into a power rack. Place the balls of your feet on a step or high block. Lock your knees and keep them locked throughout the movement. Keep your toes pointed straight forward. Raise your heels smoothly as high as possible and try to stand on your big toes. Pause. Lower your heels slowly until you feel a deep stretch in your calves. Repeat the slow raising and lowering.

● LATERAL RAISE WITH DUMB-BELLS: This is one of the very best exercises for your deltoids. Use light dumbbells and keep the movement very strict. With the dumbbells in your hands and your elbows locked, raise your arms sideways until they are slightly above the horizontal. Pause in the top position. Make sure your palms are facing down. Lower the dumbbells slowly to your sides. Repeat.

● OVERHEAD PRESS: In a standing position, place a barbell in front of your shoulders. Your hands should be shoulder-width apart. Press the barbell smoothly overhead. Do not cheat by bending your legs or arching your back. Lower the barbell slowly to your shoulders and repeat.

● NAUTILUS PULLOVER: No barbell or dumbbell exercise works your lats as thoroughly as the Nautilus pullover machine. (If you do not have access to a pullover machine, you'll have to make do with the barbell pullover on a bench.) Sit in the Nautilus pullover machine. Adjust the seat until the tops of your shoulders are in line with the axis of rotation of the movement arm. Fasten your seat belt. Leg-press the food pedal and place both elbows on the pads by your ears. Remove your feet. Rotate your elbows back behind your head until your lats are in a comfortable stretch. Pull the movement arm forward and down until the horizontal bar touches your midsection. Pause. Move slowly back to the stretched position and repeat.

● PULLDOWN BEHIND NECK: You'll need a lat machine for this exercise. Grasp the overhead bar with your palms up and your hands slightly wider than your shoulders. Anchor

Triceps extension: Keep your elbows in a vertical position.

yourself in a seated position. Pull the bar smoothly behind your neck. Pause. Move slowly back to the stretched position and repeat.

- BENCH PRESS: Lie on your back and position your body under the bench press racks and the supported barbell. Place your hands shoulder-width apart. Lift the barbell over your chest. Your feet should be flat on the floor in a stable position. Lower the barbell slowly to your chest. Press the barbell smoothly until your arms lock. Repeat.
- BENT-OVER ROW: In a bent-over position, grasp a barbell with a shoulder-width grip. Your torso should be parallel with the floor. Pull the barbell upward until it touches your lower chest. Pause. Return slowly to the stretched position. Repeat.
- TRICEPS EXTENSION WITH ONE DUMBBELL HELD IN BOTH HANDS: This is one of the best exercises for isolating your triceps. A dumbbell is held at one end with both hands. Press the dumbbell overhead. Your elbows should be close to your ears. Bend your arms and lower the dumbbell slowly behind your neck. Do not move your elbows; only your forearms and hands should move. Press the dumbbell back to the starting position. Repeat.

- BICEPS CURL: Grasp a barbell with your palms up and your hands about shoulder-width apart. Stand erect. With your body straight, smoothly curl the barbell to your shoulders. Slowly lower and repeat.
- STIFF-LEGGED DEADLIFT: This valuable exercise is often neglected by bodybuilders. It strongly involves the lower back, buttocks, and hamstrings. A small platform or bench should be used to increase the range of movement. Stand on the platform and grasp the barbell with an under-and-over grip. Your feet should be under the bar. Lift the barbell to the standing position. With your knees locked, lower the barbell to the stretched position and smoothly lift it back to the top. Repeat.

APPLYING THE BASICS

That's it! One set of fourteen exercises. When each exercise is carried to momentary muscular failure, your muscles will be optimally stimulated and your growth will be maximized.

Your goal is to use as much weight as possible on each exercise for your required number of repetitions, as much weight as possible *in good form.*

It is also important that you adhere to the food recommendations: 3,100 calories a day for Week 1.

Remember—for massive muscles you must practice the basics.

Biceps curl: Keep your body straight as you lift the barbell. Do not lean forward or backward.

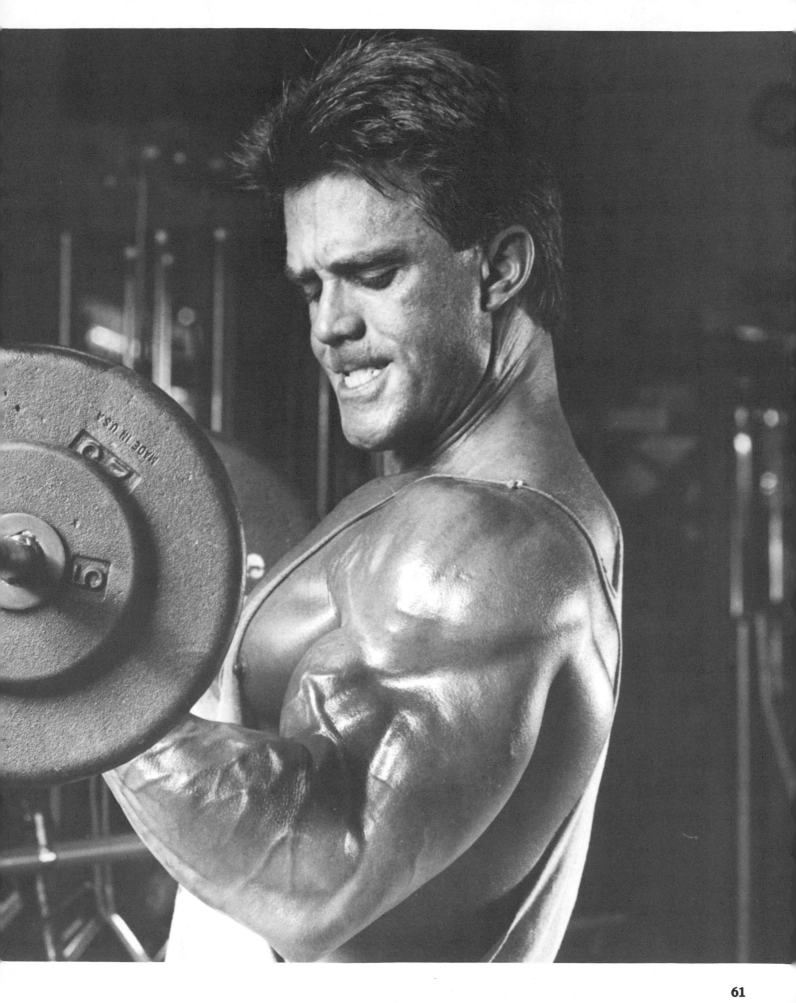

If you want your arms to grow, work your legs. This is one of the most important physiological principles that must be applied in building mass. Furthermore, it is essential that you train both your lower and upper body on the same day.

2

WEEK 2: WORKING YOUR ENTIRE BODY

Leg extension: Do not lean forward as you straighten your legs.

Few if any bodybuilders can display
the pleasing proportions of Ed Kawak.

More than seventeen years of research with athletes at Nautilus Sports/Medical Industries has shown that the human body grows most effectively when the training program is well rounded. In other words, for maximum muscular growth you must include exercises for each of your major muscle masses.

Your body functions as an entire unit. It does not function well if it is segmented or torn apart.

It should be obvious that the various types of split routines are a step in the wrong direction. It is impossible to work your upper body without affecting your lower body, or vice versa, at least to some degree. Split-routine training moves you in the direction of inefficiency. Thus you always feel you have to do more and more exercise. In a matter of weeks most bodybuilders who practice split routines are in a state of overtraining.

To build massive muscles in the most productive manner exercise your entire body each time you train. Doing so guarantees the most efficient use of your recovery ability.

Above. Leg press: Don't be afraid to go heavy on this exercise. But keep your form strict.
Below. As a variation, you can do the pullover by lying across a bench and using dumbbells instead of a barbell.

GENERAL SERVINGS AND SAMPLE MENU FOR WEEK 2

GENERAL SERVINGS (TOTAL CALORIES: 3,200 PER DAY)

Basic Food Group	Recommended Daily Servings
Meat	5.50
Milk	5.00
Fruit/Vegetable	10.00
Bread/Cereal	10.00
Other Foods	7.75

SAMPLE MENU (TOTAL CALORIES: 3,200 PER DAY)

Breakfast	Calories:	870
2 English muffins, toasted		280
3 tablespoons of jelly or preserves		150
3 eggs, any style		240
½ cantaloupe		50
1 cup of milk, whole		150

Lunch	Calories:	845
Roast beef sandwich: 5 ounces of roast beef, lean		300
1 onion roll		200
1½ ounces of cheese, swiss or cheddar		150
mustard		10
1 pickle, dill		15
½ cantaloupe		50
½ cup of peas, sweet		70
1 cup of tomato juice		50

Dinner	Calories:	1155
1 cup of orange juice, unsweetened		120
2½ ounces of baked ham, hot or cold		200
½ cup of cottage cheese, uncreamed		60
½ cup of peas, sweet		70
15 crackers, saltine		155
2 muffins, cornbread		300
1 cup of milk, whole		150
1 banana, frozen		100

Snack	Calories:	330
Chocolate milkshake: ¾ cup of milk, dry powdered		200
½ cup of water		
1½ cup of ice cubes		
2½ tablespoons of chocolate syrup, thin		130

WEEK 2: BASIC ROUTINE

	Date:	Date:	Date:
1. Leg extension			
2. Leg curl			
3. Full squat or leg press			
4. Reverse leg raise			
5. Calf raise			
6. Lateral raise			
7. Overhead press			
8. Nautilus pullover			
9. Pulldown behind neck			
10. Bench press			
11. Bent-over row			
12. Triceps extension			
13. Biceps curl			
14. Nautilus 4-way neck			

BASIC MASS-BUILDING ROUTINE FOR WEEK 2

Your basic routine for Week 2 is the same as Week 1, with the last exercise being the lone exception: the Nautilus 4-way neck machine is substituted for the stiff-legged deadlift.

1. Leg extension
2. Leg curl
3. Full squat or leg press
4. Reverse leg raise
5. Calf raise
6. Lateral raise with dumbbells
7. Overhead press
8. Nautilus pullover
9. Pulldown behind neck
10. Bench press
11. Bent-over row
12. Triceps extension with one dumbbell held in both hands
13. Biceps curl
14. Nautilus 4-way neck

Review the instructions for exercises 1–13 from the previous chapter. Perform each exercise in the same manner. The last exercise, the Nautilus 4-way neck, is done according to the instructions below. If a 4-way neck machine is not available, you can do similar exercises with a neck harness.

Nautilus 4-way neck machine: This machine provides comfortable, safe, direct exercise for the important neck muscles in four directions: back, front, left side, and right side.

Back extension: Sit in the machine with the back of your head next to the movement arm. Adjust your seat height until the back of your head is squarely in the middle of the pads when you are seated exactly. Extend your head as far back as possible. Pause. Return slowly to the stretched position and repeat. Your guideline number of repetitions for your neck should be the same as your arms.

Above. 4-way neck, back extension: Move your head backward against the pads.
Below. 4-way neck, front flexion: Always start the movement smoothly, without jerking.

Front flexion: Turn and face the machine. Your nose should now be in the center of the pads. Stabilize your torso by lightly grasping the handles. Move your head smoothly toward your chest. Pause. Return slowly to the stretched position. Repeat.

Lateral contraction: Turn your body in the machine until your left ear is in the center of the pads. Stabilize your torso by lightly grasping the handles. Move your head toward your left shoulder. Pause. Keep your shoulders square. Return slowly to the stretched position. Repeat. Turn in the machine until your right ear is in the center of the pads and work the right side of your neck in the same manner.

CHARTING YOUR PROGRESS

At the end of Week 2, you should be from 5 to 10 percent stronger on all of your exercises. For example, on the leg extension machine, initially you may have done 100 pounds for ten repetitions, or 100/10. One week later using the progression guidelines, you

did 105/10. Two weeks later, you performed 110/10. Thus, going from 100 pounds to 110 pounds, with the same number of repetitions, means you are 10 percent stronger in the leg extension. Your goal is to progress in a similar manner in all your exercises.

If you are not from 5 to 10 percent stronger in most of your exercises at the end of Week 2, then you must very simply *work harder!*

At the end of Week 2, your body weight should be on the rise. Most young bodybuilders should note an average weight gain of 2 to 4 pounds for the first two weeks. The best time to weigh yourself is immediately before your workouts. Strip down to your bare essentials and be consistent with your weighing techniques. Record your weight on your workout card.

If you've gained less than 2 pounds, ask yourself several questions:
1. Are you following the routines exactly as directed? If not, you may be neglecting an important exercise or body part.

2. Are you being pushed to momentary muscular failure on all your exercises? If not, then your muscles are not being stimulated to grow in the most efficient manner.

3. Are you getting adequate rest and relaxation between your workouts? For building maximum muscle mass, it is a good idea to keep other fitness activities — such as jogging, racquetball, basketball, and water skiing — to a bare minimum.

4. Are you eating the recommended servings each day from the Basic Four Food Groups? If not, you may be getting an inadequate ratio of muscle-building calories.

On the other hand, if you gained more than 4 pounds during the first two weeks you may want to double-check the number of calories that you are consuming each day. Too many calories will lead to a gain in fat, not muscle. You can make sure the weight you gain is muscle by monitoring your percentage of body fat on a weekly basis.

Harder exercise is the secret to maximum growth stimulation.

3

WEEK 3: STRESSING YOUR THIGHS

Eddie was wrestling with a dilemma. Fortunately, it wasn't one of life's toughest problems. It was more like trying to choose between dating a Dallas Cowboy cheerleader and a Miss America finalist. In this instance the situation involved not girls but the goal Eddie wanted to set for putting muscle onto his thighs.

"I know Tom Platz has the greatest thighs in bodybuilding, and I'm impressed by their size and shape," Eddie said prior to his first workout for Week 3. "But in the back of my mind I can still see those thighs of Casey's on his bicycle as he pedaled by the school yard. Casey's massive thighs have always been my ideal."

"Platz's thighs are almost too superhuman," I said, trying to piece together the sentiment Eddie was expressing. "It's difficult to conceive of any bodybuilder having a realistic goal of attaining the thighs of Tom Platz."

Eddie agreed. "I know my thighs will never look like Platz's. I'd certainly settle for Casey's."

I smiled to myself. It would be easy to organize a thigh program Eddie could believe in.

Above. Kawak's legs look great from all angles. *Opposite.* Size, shape, and muscularity describe the thighs of Ed Kawak.

Leg extension: Do as many strict repetitions as you possibly can.

GENERAL SERVINGS (TOTAL CALORIES: 3,300 PER DAY)

Basic Food Group	Recommended Daily Servings
Meat	5.00
Milk	5.25
Fruit/Vegetable	11.00
Bread/Cereal	11.00
Other Foods	8.00

SAMPLE MENU (TOTAL CALORIES: 3,300 PER DAY)

Breakfast	Calories: 795
3 ounces of ham slices, lean, packaged	185
1 English muffin, toasted	140
2 tablespoons of jelly or preserves	100
1 cup of yogurt, low-fat, flavored	200
¾ cup Cheerios	70
¼ cup of raisins	100

Lunch	Calories: 880
Cheeseburger: 5 ounces of ground beef	300
1 ounce of cheese, swiss or cheddar	100
1 hamburger roll	200
½ cup of noodles, cooked	80
2 teaspoons of butter or margarine	70
2 carrots, sliced	40
1 cup of broccoli, steamed, with lemon juice	40
½ cup of applesause, unsweetened, sprinkled with cinnamon	50

Dinner	Calories: 1250
⅓ cup of orange juice, unsweetened	35
3 fried chicken legs, 2 ounce size	300
1 cup of rice, white	180
3 slices of bread	240
3 teaspoons of butter or margarine	105
1 cup of broccoli, steamed, with lemon juice	40
1 cup of corn, cream-style	200
1 cup of milk, whole	150

Snack	Calories: 380
Banana milkshake: 2 cups of milk, skim	180
½ cup of ice milk, vanilla	100
1 banana, frozen	100

WEEK 3: THIGH ROUTINE

	Date:	Date:	Date:
1. Leg press			
2. Leg extension			
3. Full squat			
4. Reverse leg raise			
5. Leg curl			
6. Stiff-legged deadlift			
7. Lateral raise			
8. Overhead press			
9. Nautilus pullover			
10. Bench press			
11. Biceps curl			
12. Triceps extension			
13. Chin-up			
14. Dip			

"Now that we've decided that you want your legs to look like Casey's," I said, "I'm going to pattern your thigh routine after one I saw him do frequently. It will make you weak-kneed, I promise. And best of all, it's guaranteed to build massive thighs."

DOUBLE PRE-EXHAUSTION FOR YOUR THIGHS

Normal pre-exhaustion training is performed when a single-joint exercise is immediately followed by a multiple-joint exercise. For example, the leg extension is followed by the leg press. The leg extension involves the muscles of the knee joint only. The leg press works the muscles surrounding the hip, knee, and ankle joints. The leg extension therefore pre-exhausts the quadriceps. Then, before the frontal thigh can recover, the leg press brings into action the gluteals, hamstrings, and gastrocnemius to force the quadriceps to a deeper state of exhaustion.

Double pre-exhaustion for the thigh goes a step further. Instead of performing two exercises back-to-back, you do three in a row. The best order for a double pre-exhaustion cycle is multiple-joint, single-joint, and multiple-joint.

Casey Viator frequently used double pre-exhaustion to stress his massive thighs into additional growth. It worked for him and it will work for you.

The first three exercises involve double pre-exhaustion for your quadriceps. The next three apply double pre-exhaustion to your hamstrings. Exercises 7–14 are for your upper body.

1. Leg press, immediately followed by
2. Leg extension, immediately followed by
3. Full squat

4. Reverse leg raise, immediately followed by
5. Leg curl, immediately followed by
6. Stiff-legged deadlift

7. Lateral raise with dumbbell
8. Overhead press
9. Nautilus pullover
10. Bench press
11. Biceps curl
12. Triceps extension with one dumbbell held in both hands

13. Chin-up
14. Dip

● LEG PRESS: Although the leg press involves several muscles, the primary emphasis is on the quadriceps. Don't be afraid to work heavy. Lift and lower the resistance smoothly and slowly. After the final repetition, it is important that you move quickly from the leg press to the leg extension. As short as three seconds between these two machines will allow significant recovery of your quadriceps. Thus it is imperative on any pre-exhaustion cycle to move from one exercise to the next in *less than three seconds.*

● LEG EXTENSION: Your quadriceps will be partially exhausted from the leg press, so you'll have to reduce the resistance that you normally handle on the leg extension. Immediately get into the leg extension and start performing the exercise. Try to keep your head against the seat back, relax your face and neck, and avoid excessive gripping of the handles. Be sure to pause for a second in the top position, but do not pause at the bottom end of the movement. Even though your quadriceps will be screaming for mercy, grind out as many repetitions as possible. On the last repetition, your training partner should assist you in running to the squat racks. Granted, you won't feel like running, but do the best you can. Once again, it is important to move from the second to the third exercise in less than three seconds.

● FULL SQUAT: You'll have to cut the weight that you normally would use for squats by approximately 50 percent. Get into position with the barbell on your shoulders and start squatting. The idea is to bring into action your buttocks and hamstrings to force your almost exhausted quadriceps to a level of fatigue that you have never before experienced. It's this deeper level of fatigue that shocks your thighs into massive growth.

Do your limit number of repetitions on the squat and take a several-minute break to get your breathing back to normal. You'll need the rest for your hamstrings cycle.

● REVERSE LEG RAISE: This exercise involves the secondary function of

Full squat: Use an alert spotter for
this exercise and grind out the repetitions.

Leg curl: Do not keep your hips flat on the pad in the contracted position. Lifting your hips slightly relaxes your quadriceps so you can better work your hamstrings.

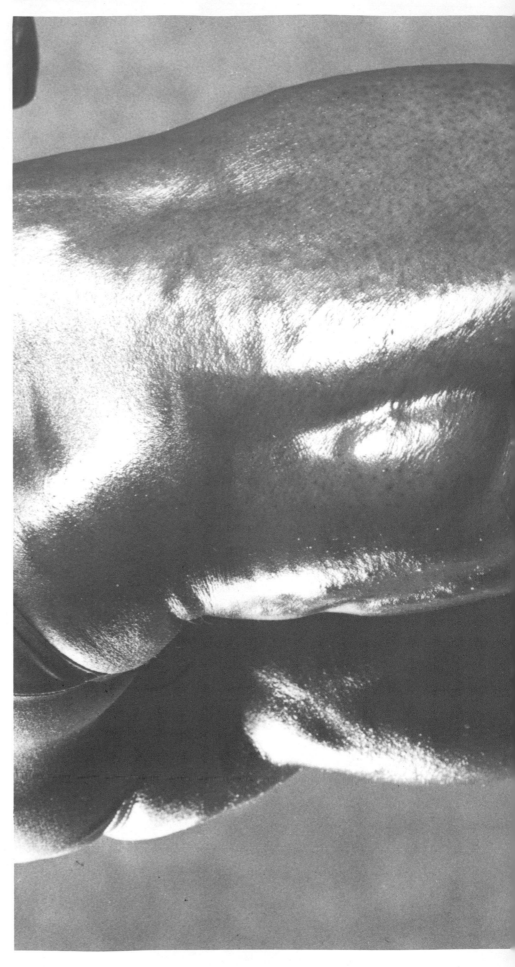

your hamstrings and the primary function of your gluteus maximus. Lie face down. Raise your thighs backward, pause in the top position, and squeeze your buttocks together. Lower to the floor and repeat for maximum repetitions. After the final repetition, instantly get into the leg curl machine.

• LEG CURL: Use approximately 10 percent less on this exercise than you would normally employ. Curl the movement arm, pause in the contracted position, and lower slowly. Continue until you can no longer lift the movement arm. Move quickly to the deadlift bar.

• STIFF-LEGGED DEADLIFT: The deadlift brings into action your buttocks and upper body to force your pre-exhausted hamstrings into growing. Concentrate on a smooth lifting and lowering movement of the barbell. Be sure to keep your knees locked on all repetitions, with the exception of the first lifting movement. Also, standing on a raised platform will increase your range of motion with the barbell.

After the last repetition, you'll need another rest period before working your upper body.

Upper body exercises 7–12 are performed in the same manner as previously described in Week 1. Only the last two exercises, the chin-up and the dip, require direction.

• CHIN-UP: Grasp the horizontal bar with an underhand grip, and hang. Your hands should be shoulder-width apart. Pull your body upward so your chin is over the bar. In fact, try to touch the bar to your chest. Pause. Lower your body slowly to the hanging position. Repeat.

• DIP: Mount the parallel bars and extend your arms. Bend your arms and lower your body slowly. Stretch at the bottom and recover smoothly to the top position. Repeat.

THE MOST PRODUCTIVE CYCLE

This double pre-exhaustion cycle for your quadriceps and hamstrings is the most demanding in the book. And because it is the most demanding, it is the most productive. I won't be surprised if you add at least an inch to the circumference measurements of each thigh. Apply yourself during the third week and you won't be disappointed.

The massively developed hamstrings of Wilf Sylvester.

WEEK 4:
PUMPING
YOUR CALVES

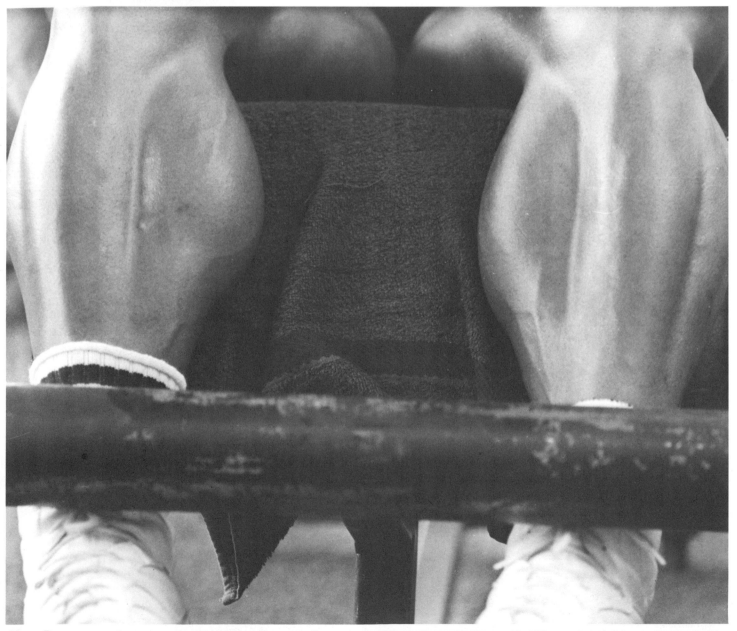

Above. Proper exercise can pump your calves by more than one-half inch. *Opposite*. Danny Padilla contracts his muscular calves.

GENERAL SERVINGS AND SAMPLE MENU FOR WEEK 4

GENERAL SERVINGS (TOTAL CALORIES: 3,400 PER DAY)

Basic Food Group	Recommended Daily Servings
Meat	5.50
Milk	5.50
Fruit/Vegetable	11.00
Bread/Cereal	11.00
Other Foods	8.25

SAMPLE MENU (TOTAL CALORIES: 3,400 PER DAY)

Breakfast	Calories:	940
1 cup of orange juice, unsweetened		120
4 waffles, frozen		480
2 tablespoons of syrup		105
2 slices of Canadian bacon, crisp		100
1½ cups of milk, skim		135

Lunch	Calories:	995
Tuna Salad: 1 6½-ounce can of tuna, packed in water		200
1 tablespoon of mayonnaise		100
2 eggs, boiled and diced		160
1 tomato, sliced		35
relish		5
Combine all ingredients on: 3 lettuce leaves		10
2 slices of bread		160
1 apple		80
20 pretzels, small stick type		70
1½ cups of milk, skim		135

Dinner	Calories:	1205
Spaghetti & meat sauce: 4 ounces of ground lean beef, sauteed		605
with desired vegetables as mushrooms, peppers, and onions; add		
desired spices; add 1 cup of tomato sauce, 1 cup of spaghetti, cooked		
2 tablespoons of Parmesan cheese		70
9 breadsticks		335
1 teaspoon of butter or margarine		35
Spinach salad: 2 cups of spinach, torn		40
1 cup of mushrooms, sliced		20
1¼ teaspoons of Italian dressing		10
1 cup of milk, powdered or skim		90

Snack	Calories:	270
Orange smoothie: ½ cup of dry powdered milk		105
¾ cup of ice cubes		
1 cup of orange juice, unsweetened		120
¼ teaspoon of vanilla		
1 tablespoon of sugar		45

WEEK 4: CALF ROUTINE

	Date:	Date:	Date:
1. Calf raise			
2. Leg Curl			
3. Calf raise on leg press			
4. Leg curl			
5. Overhead press			
6. Chin-up			
7. Bench press			
8. Pulldown behind neck			
9. Incline press			
10. Shoulder shrug			
11. Trunk curl			
12. Side bend			
13. Leg extension			
14. Leg press			

When muscles are engaged in any kind of work, they demand an increased blood flow, or circulation. Increased circulation is important for two reasons. First, it provides the muscles with the fuel and nutrients they require. Second, the flowing blood picks up and removes the larger-than-normal amount of waste that is being produced.

If the work continues for some time, the flow of blood into the muscles and subsequent outflow strike a happy balance. A feature of this balance is a slight enlargement of the working muscles.

However, if the muscles are worked at maximum intensity, as in a heavy barbell exercise, and if the repetitions are executed consecutively, the muscles produce congestion in their interior, thereby swelling the involved body part to a degree that is often astonishing. The process by which the muscles have become engorged with fluids is called *pumping.* A pumped upper arm or calf may gain a full half-inch over its normal size.

When pumped to that extent, an arm or calf will feel very heavy, which is not surprising, since its actual weight has temporarily been increased. It will also feel stiff, since its flexibility will be temporarily reduced. In most cases the degree of muscularity will be decreased. The muscles will look much larger and will be much larger but will appear less defined than they normally do. In a few cases, particularly in a bodybuilder with an extreme degree of muscularity, a pumped muscle may actually appear more defined.

In most forms of everyday activity the effects of pumping usually occur without being noticed. For example, few people are aware that their calves are usually a quarter-inch larger at night than they are early in the morning. Their calves increase in size during the day as a result of the pumping that occurs naturally from normal standing and walking. During the night, when the calves are resting, their circulation requirements are reduced greatly, and their size is decreased.

The pumped calves of Brian House.

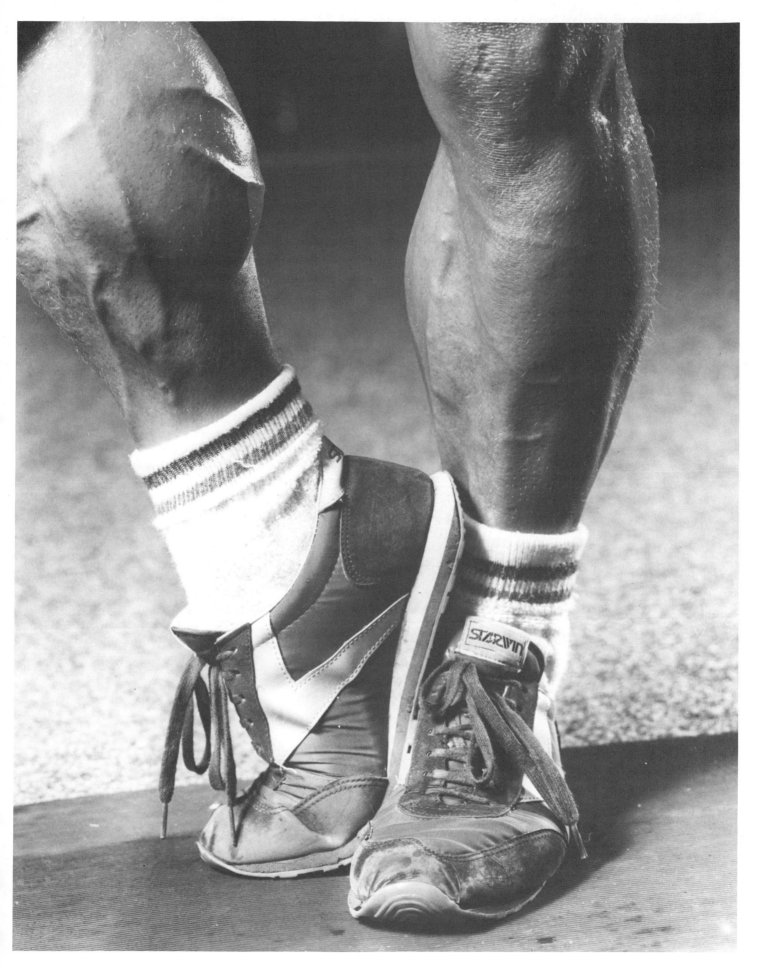

PUMPING AND FUTURE GROWTH

Above-average muscular pumping is usually an indication that growth stimulation is occurring. If no noticeable degree of pumping is produced, then an exercise is doing little to build muscular size and strength. Although a noticeable degree of pumping becomes evident during any really productive exercise, it does not follow that an extreme degree of pumping is a sign of doing proper exercise. Actually, it is possible to produce an extreme pump from exercises that do nothing to build either size or strength.

Light movements performed in sets of very high repetitions, especially if the movements are restricted in range, will produce muscular pumping without building size or strength. On the other hand, several sets of heavy, full-range repetitions, carried to momentary muscular failure, will produce the same degree of pumping, and they will also induce growth stimulation.

Your pumped calf measurement will give you an advance indication of your future development in that body part. If your calf normally pumps one-half inch during a specific routine, and it abruptly shows a gain of three-quarters of an inch as a result of the same routine, your calf is ready to grow during the next forty-eight hours.

The ability to pump your calf to a particular size usually precedes the actual growth of that muscle to the same size as its earlier pumped-up measurement. In other words, if you pump your calf until it is larger than

Above left. Calf raise: Keep your knees locked throughout the movement.

Left. Leg curl: Pause in the top position.

Opposite, above. Another good exercise for your lower legs is the seated calf raise.
Opposite, below. Incline press: Do not bounce bar off your chest. Perform the movement smoothly.

normal at the end of a specific cycle, you can be assured that your calf has the solid capacity to increase in size until it is as large as it was when you first pumped it to its new dimension.

This chapter contains a routine that is sure to pump your calves to the extreme. In doing so, you'll be bombarding them with maximum growth stimulation.

CALF-PUMPING ROUTINE

The body part that is featured each new week is worked first. Four exercises compose the calf cycle. It is important to perform 1 and 2 back-to-back without rest, and 3 and 4 in the same pre-exhaustion fashion.

Let's examine the four calf exercises, as well as the other ten movements that make up your routine for Week 4.

1. Calf raise on multi-exercise, immediately followed by
2. Leg curl

3. Calf raise on leg press, immediately followed by
4. Leg curl

5. Overhead press
6. Chin-up
7. Bench press
8. Pulldown behind neck
9. Incline press
10. Shoulder shrug
11. Trunk curl
12. Side bend
13. Leg extension
14. Leg press

● CALF RAISE ON MULTI-EXERCISE: You'll need a Nautilus multi-exercise machine or some type of calf machine for this exercise. Adjust the belt on the multi-exercise machine comfortably around your hips. Place the balls of your feet on the first step and your hands on the front of the carriage. Lock your knees and keep them locked throughout the movement. Elevate your heels as high as possible and try to stand on your tiptoes. Pause in the highest position. Lower your heels slowly. Stretch at the bottom and repeat. After the final repetition, instantly move to the leg curl.

● LEG CURL: The calf raise has effectively pre-exhausted your gastrocnemius muscles. Since your gastrocnemius muscles also cross the back of your knee, their secondary function is to assist the hamstrings in bending the leg. Thus, by doing the leg curl immediately after the calf raise, you'll be able to focus the involvement on your calves. You'll also have to use approximately 10 percent less resistance on the machine than you would normally handle. Do as many leg curls as you can in good form. Rest exactly thirty seconds and move to the leg press machine for more calf raises.

● CALF RAISE ON LEG PRESS: The leg press provides a slightly different feel than the standing calf raise. Leg press the movement arm until your knees are locked. Make sure the balls of your feet are in solid contact with the foot board. Perform the calf raise by extending your feet. Pause in the contracted position. Move your feet backward until a deep stretch is felt in your calves. Do not bend your legs. Keep your knees locked throughout the movement. Repeat until momentary muscular failure. Quickly exit the leg press and move back to the leg curl.

● LEG CURL: Your calves should really be burning now. This final set of leg curls will add the finishing touch to your engorged lower legs. Reduce the weight on the leg curl by another 10 percent. Curl the pads to your buttocks for as many repetitions as possible. Avoid jerking movements. Make your calves feel the involvement clear to the bone. After your final repetition, get up and walk around shaking your feet for several minutes. The pumping in your calves should be stunning.

Because of the temporary congestion in your calves, your other two leg exercises for this routine will be saved for last. Exercises 5–8 have been previously detailed in other chapters, so I'll move directly to the incline press.

● INCLINE PRESS: You'll need a forty-five-degree incline bench for this exercise. Stabilize the barbell over your upper chest with your arms extended. Your hands should be slightly wider than your shoulders. Lower the barbell slowly to your upper chest. Press it back to the extended position. Repeat.

● SHOULDER SHRUG: The shoulder shrug is a great exercise for your upper back and neck. Grasp a heavy barbell with an under-and-over grip. Stand with the barbell hanging at arm's length. Shrug your shoulders and try to touch them to your ears. Pause. Lower slowly to the bottom. Repeat.

● TRUNK CURL: This exercise strongly activates your abdominal muscles. Lie face up on the floor with your hands behind your head. Keep your chin on your chest. Bring your heels up close to your buttocks and spread your knees. Do not anchor your feet under anything, and don't have a partner hold your knees down. The idea is to work your abdominals without involving your hip flexors. Try to curl your trunk smoothly to a sitting position. Only one-third of a standard sit-up can be performed in this fashion. Pause in the contracted position and lower your trunk slowly to the floor. Repeat. As this exercise becomes easier, hold a barbell plate behind your neck.

● SIDE BEND: Grasp a heavy dumbbell in your right hand. Stand erect and place your left hand on the top of your head. Bend laterally to your right. This bending stretches your left obliques. Return smoothly to the erect position. Repeat for the correct number of repetitions. Switch the dumbbell to your left hand and perform the same number of side bends to your left side.

Your last two exercises for Week 4 are the leg extension and the leg press. Perform each exercise to momentary muscular failure in good form.

DEALING WITH SORENESS

If you haven't worked your lower legs hard in the past several months, the routine in this chapter will probably make your calves very sore. Soreness is a result of many factors, such as pumping, stretching, and pre-exhausting. Do not be alarmed, however. This is a natural reaction. The pain will be gone in several days. Furthermore, exercising your calves helps to eliminate the soreness faster.

Soreness is also an indication that the routine has stimulated your calves to grow. And grow they will if you perform the routine exactly as directed.

Ian Lawrence displays a balanced upper and lower body.

5

WEEK 5:
BROADENING
YOUR SHOULDERS

At the beginning of Week 5 Eddie stated a priority.

"I'd rather have four inches on my shoulders than four inches on my arms," he said.

My expression indicated to him that I was curious.

"Because lots of times you can't wear T-shirts and tank tops. Sometimes you have to wear shirts with sleeves," Eddie explained.

I realized his point but I wanted to hear him say it. "What's wrong with dressing up occasionally?" I asked.

"Nothing, except your biceps and triceps don't get a chance to impress people," Eddie replied. "You know what I mean?"

Affirmative.

"But it's different with broad shoulders," Eddie continued. "Even when you have to wear a long-sleeve shirt or coat your shoulders are still there, making a strong statement."

Eddie was perusing a muscle magazine featuring Frank Richard. "I'd sure like to have shoulders like his," he said, thumbing through Chris Lund's *Bodybuilding Monthly*, which is published in Great Britain.

"Let me see," I said, looking over Eddie's shoulder at the photos of Richard.

"He's got some broad shoulders all right — but they're not as wide as Scott Wilson's. Scott has the most

massive shoulders I've ever seen on a man."

"Then I want shoulders as broad as Scott Wilson's," Eddie replied.

"That's an excellent goal to have, Eddie ole boy," I concurred. "And I've got just the shoulder routine that will make Richard and Wilson green with envy when they see your results."

The overhead press is a major shoulder exercise that has been used by every champion bodybuilder.

The outstanding shoulders of Frank Richard.

GENERAL SERVINGS AND SAMPLE MENU FOR WEEK 5

GENERAL SERVINGS (TOTAL CALORIES: 3,500 PER DAY)

Basic Food Group	Recommended Daily Servings
Meat	5.50
Milk	5.50
Fruit/Vegetable	12.00
Bread/Cereal	11.00
Other Foods	8.50

SAMPLE MENU (TOTAL CALORIES: 3,500 PER DAY)

Breakfast	Calories:	795
1 cup of apple juice, unsweetened		120
Grilled cheese sandwich: 2 slices of bread		160
1½ ounces of cheese, swiss or cheddar		150
1 teaspoon of butter or margarine		35
3 ounces of sausage, brown and serve		240
1 cup of milk, skim		90

Lunch	Calories:	1100
Pineapple chicken salad: 1 cup of chicken, no skin, diced; may used canned, drain broth		480
½ cup of celery, diced		10
1 apple, diced		80
¼ cup of raisins		100
½ cup of pineapple, chunk, unsweetened		50
3 tablespoons of sour cream		70
Combine all ingredients on: 6 lettuce leaves		20
1 bagel		140
4 teaspoons of jelly or preserves		70
1 cup of milk, skim		90

Dinner	Calories:	1125
4 ounces of pork chop, with bone		150
1 cup of rice, brown		200
½ cup of baked beans		140
3 dinner rolls		300
3½ teaspoons of butter or margarine		120
1¾ cups of apple juice		215

Snack	Calories:	470
Pineapple milkshake: 2 cups of milk, skim		180
½ cup of ice milk, vanilla		100
½ cup of pineapple, chunk, unsweetened		50
40 pretzels, small stick type		140

WEEK 5: SHOULDER ROUTINE

	Date:	Date:	Date:
1. Lateral raise			
2. Overhead press			
3. Shoulder shrug			
4. Upright row			
5. Bent-over raise			
6. Press behind neck			
7. Full squat			
8. Calf raise			
9. Nautilus pullover			
10. Bench press			
11. Biceps curl			
12. Triceps extension			
13. Wrist curl			
14. Reverse wrist curl			

ROUTINE FOR BROAD SHOULDERS

Six exercises make up the shoulder cycle, and eight exercises are for the other body parts. The shoulder cycle, like previous chapters, uses the pre-exhaustion principle.

1. Lateral raise with dumbbells, immediately followed by
2. Overhead press

3. Shoulder shrug, immediately followed by
4. Upright row

5. Bent-over raise with dumbbells, immediately followed by
6. Press behind neck

7. Full squat
8. Calf raise
9. Nautilus pullover
10. Bench press
11. Biceps curl
12. Triceps extension with one dumbbell held in both hands
13. Wrist curl
14. Reverse wrist curl

● LATERAL RAISE WITH DUMB-BELLS: The lateral raise pre-exhausts your deltoids for the following exercise. Perform as many strict repetitions as possible. Move quickly to the overhead press. Remember, the time between exercises in a pre-exhaustion cycle must be three seconds or less.

● OVERHEAD PRESS: Press the barbell over your head smoothly. Do not rest in the lockout. Lower slowly to your shoulders and repeat the pressing. You're permitted to cheat a little on the last repetitions of this exercise, but keep it to a minimum. Rest for about sixty seconds and ready yourself for the shoulder shrug and upright row.

● SHOULDER SHRUG: You'll be able to shoulder shrug more than you can upright row. To facilitate the performance of these two exercises back-to-back, load the barbell such that all you have to do is place the barbell on the floor and strip off a few plates on both sides. Grasp the barbell with an under-and-over grip and stand erect. Shrug your shoulders smoothly as high as possible. Lower slowly to the stretched position. Repeat for maximum repetitions. Immediately place the barbell on the floor, remove the

Scott Wilson has the broadest shoulders in professional bodybuilding.

necessary plates from both sides, pick the barbell back up, and begin the upright row.

• UPRIGHT ROW: Grasp the barbell with an overhanded grip and stand erect. Your hands should be about six inches apart. Pull the barbell smoothly to your neck. Keep your elbows higher than your hands. Pause. Lower slowly and repeat until you are unable to get the bar to your neck. Rest for about sixty seconds before going to the next two exercises.

• BENT-OVER RAISE WITH DUMBBELLS: This is a great exercise for your posterior deltoids, which are neglected by many bodybuilders. With a pair of light dumbbells in your hands, bend over until your torso is parallel to the floor. Let the dumbbells hang. Raise the dumbbells backward as far as possible. Do not bend your elbows. Keep them locked solidly throughout the movement. Pause in the top position. Lower slowly and repeat. After your last repetition, move quickly to the press behind neck.

• PRESS BEHIND NECK: Your shoulders should be pretty well exhausted as you start to do this exercise. Now is the time to call upon the strength of your triceps to squeeze a bit more involvement out of your deltoids. In a standing position, place the barbell behind your neck. Your hands should be about six inches wider than in the overhead press. Press the barbell in a strict manner over your head. Lower slowly behind your neck. Grind out as many repetitions as possible.

Your shoulders should now be pumped to the explosion point. That's good because new growth is being stimulated in previously unused fibers.

Rest by walking around, shaking your arms, and drinking water. In several minutes you'll be ready to work the rest of your body.

Exercises 7–12 have been described previously. Do them in the listed order. The last two exercises should be done in the following manner.

• WRIST CURL: Grasp a barbell with a palms-up grip. Rest your forearms on your thighs and the back of your hands over your knees, and sit on the end of a bench. Lean forward until the angle between your upper arms and

Nicole Bass's wide shoulders add to her pleasing proportions.

forearms is less than 90 degrees. This allows you to isolate your forearms more completely. Curl your hands smoothly and contract your forearm muscles. Pause, and lower the barbell slowly. Repeat.

● REVERSE WRIST CURL: You'll have to reduce the weight considerably on this exercise. Assume the same position as for the wrist curl, except reverse your grip. Move the backs of your hands upward. Lower slowly and repeat.

YOU CAN'T HIDE BROAD SHOULDERS

Apply the shoulder routine for three consecutive workouts. In just one week, you'll feel and see the difference that high-intensity training makes on your deltoids.

And remember, you can't hide broad shoulders!

CHECKING YOUR PROGRESS

You have now completed five weeks of the ten-week program. Before getting into Week 6, let's check your progress at this the halfway point.

At the end of Week 5, you should be from 15 to 25 percent stronger in your basic exercises, such as the leg extension, pullover, bench press, and biceps curl. These four exercises are included in almost every chapter, and they are the best ones to use for checking your progress. If you are not from 15 to 25 percent stronger in these four exercises, once again you need to work harder. Also, you must eliminate any intervening activities that you are doing between your workouts that may be limiting your ability to recover.

If you've practiced the dietary recommendations so far, you should note a body-weight gain of from 5 to 10 pounds. If you do not fall into this range, please review the questions on page 67.

Left. Lateral raise: Keep your knuckles up on this movement.

Opposite. Overhead press: Use your triceps to force your deltoids to a deeper level of exhaustion.

Above. Penny Price has not neglected her deltoids. *Opposite.* Press behind neck: Do not rest in the lockout. Immediately lower the barbell slowly behind your neck and repeat the process.

Above. Wrist curl: Come to a complete stop in the contracted position. *Opposite.* John Terilli is noted for his broad shoulders.

6

WEEK 6: POWERIZING YOUR BACK

"His back is unreal," Eddie whispered as we watched Mike Quinn pose for the camera of Chris Lund. "His lat spread is almost as impressive as Tony Pearson's."

"That's not an exaggeration, Eddie," I said as Quinn muscled through his poses. "Boy, he's sure got a powerful-looking back."

Only a month before, Mike had won the 1986 Mr. Florida contest, which qualified him for the National Championships. We both thought he had a good chance of winning.

Mike Quinn's wide, powerful back did not occur by accident. In fact, two of his favorite back exercises are the Nautilus pullover followed by the chin-up.

Eddie's back cycle for Week 6, therefore, included a double dose of pullovers and chin-ups.

Above. Originally from Massachusetts, Mike recently moved to Florida and is in hard training for the National Physique Championship. *Opposite.* The powerful back of Mike Quinn. In 1984 Mike won the NABBA Mr. Universe.

GENERAL SERVINGS AND SAMPLE MENU FOR WEEK 6

GENERAL SERVINGS (TOTAL CALORIES: 3,600 PER DAY)

Basic Food Group	Recommended Daily Servings
Meat	6.00
Milk	6.00
Fruit/Vegetable	11.00
Bread/Cereal	11.00
Other Foods	8.50

SAMPLE MENU (TOTAL CALORIES: 3,600 PER DAY)

Breakfast	Calories:	850
1 cup of grapefruit juice, unsweetened		100
3 cups of cornflakes		240
1 cup of milk, skim		90
½ cup of prunes, sweetened, with pits		280
1 bagel		140

Lunch	Calories:	890
Shrimp cocktail: 7 ounces of peeled shrimp, fresh or frozen		200
½ cup of cocktail sauce		160
7 crackers, saltine		70
3 teaspoons of butter or margarine		105
½ cup of cottage cheese, uncreamed		55
20 french fries, 2-inch-long		300

Dinner	Calories:	1330
6 ounces of fried calf's liver		415
2½ tablespoons of flour		70
2 teaspoons of butter or margarine		70
1 cup of onions, cooked		40
1½ cups of mashed potatoes		200
1 tablespoon of butter or margarine		105
4 breadsticks		140
½ cup of dried apricots		200
1 cup of milk, skim		90

Snack	Calories:	530
Wheat germ shake: 2 cups of milk, skim		180
1 cup of ice milk, vanilla		200
1 egg, raw		80
¼ cup of wheat germ		70

WEEK 6: BACK ROUTINE

	Date:	Date:	Date:
1. Bent-over row			
2. Nautilus pullover			
3. Chin-up, negative only			
4. Pulldown behind neck			
5. Nautilus pullover			
6. Behind-neck chin up, negative only			
7. Leg extension			
8. Leg curl			
9. Reverse leg raise			
10. Overhead press			
11. Stiff-legged deadlift			
12. Trunk curl			
13. 4-way neck			
14. Shoulder shrug			

BACK ROUTINE

Two double pre-exhaustion cycles attack your back from all angles.

1. Bent-over row, immediately followed by
2. Nautilus pullover, immediately followed by
3. Chin-up, negative only

4. Pulldown behind neck, immediately followed by
5. Nautilus pullover, immediately followed by
6. Behind-neck chin-up, negative only

7. Leg extension
8. Leg curl
9. Reverse leg raise
10. Overhead press
11. Stiff-legged deadlift
12. Trunk curl
13. 4-way neck
14. Shoulder shrug

● BENT-OVER ROW: In a bent-over position, grasp a barbell with a shoulder-width grip. Your torso should be parallel with the floor. Pull the barbell upward until it touches your lower chest. Pause at the top. Return slowly to the stretched position. Repeat for maximum repetitions. Move quickly to the pullover.

● NAUTILUS PULLOVER: This exercise works your lats without involving the muscles of your arms. Make sure your shoulders are in line with the axis of rotation of the movement arm. Rotate the movement arm smoothly into the contracted position. Pause. Allow your arms to return until they are behind your head and your lats are fully stretched. Continue the full-range repetitions until the working muscles are exhausted. By now your lats should be aching, but your arms should be recovered from the first exercise. You'll need your arm strength to do the chin-up.

● CHIN-UP, NEGATIVE ONLY: The ideal way to do negative-only chin-ups is with the Nautilus multi-exercise machine. If you don't have access to this machine, you'll have to improvise by placing a sturdy box underneath a horizontal bar. For negative-only chin-

Rich Gaspari's back could be used for an anatomy chart.

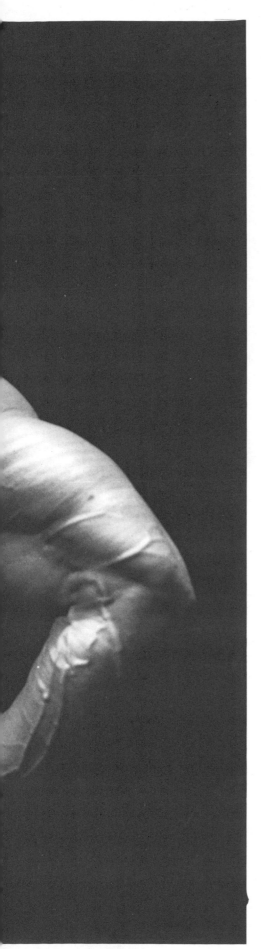

Above. Nautilus pullover: Do not short-range your repetitions. Strive for complete range of motion and you'll get better development.
Below. The leverage machine rowing exercise removes the lower back stress that many people get when they employ the barbell version.

Opposite. The wide, thick back of Scott Wilson.

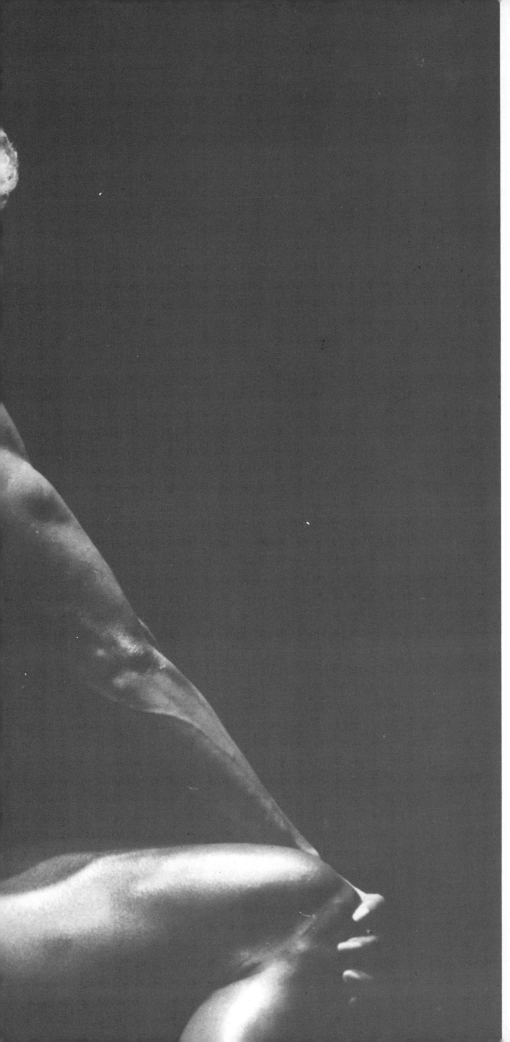

ups, the multi-exercise machine must be preadjusted. The carriage must be at the correct height and the crossbar should be in the forward position. Quickly climb the steps, place your chin over the crossbar, and bend your knees. Take eight to ten seconds to lower your body slowly, inch by inch, until your arms are completely straight. Immediately climb back to the top position and repeat the slow lowering movement. When you can perform your upper guideline number of repetitions, you should attach additional resistance to your body by using the hip belt, movement arm, and weight stack. After the negative-only chin-ups, you'll deserve a several-minute rest.

● PULLDOWN BEHIND NECK: Once again, in this second double pre-exhaustion cycle you'll be doing a multiple-joint exercise, followed by a single-joint exercise, followed by a multiple-joint, negative-only exercise. Anchor yourself securely in the lat machine. Grasp the overhead bar with your palms up and your hands slightly wider than your shoulders. Pull the bar smoothly behind your neck. Pause. Lower slowly and continue until muscular failure. Immediately exit the machine and sit in the pullover.

● NAUTILUS PULLOVER: You'll have to reduce the weight on this second set of pullovers by approximately 20 percent. Perform the first several repetitions in perfect form. Really try to isolate your lats. On your last several repetitions, loosen your form slightly. Try to get as many repetitions as you possibly can. Remove your seat belt and run to the multi-exercise machine for the chin behind neck.

● BEHIND-NECK, CHIN-UP, NEGATIVE ONLY: For this exercise you'll be using the parallel handles at the top of the multi-exercise machine. If the machine is not available to you, use a regular chinning bar, place your hands about six inches wider than your shoulders, and employ an overhanded grip. Quickly climb the steps until your

Wide, flared lats are especially important to women bodybuilders so they can compensate for the tendency that most women have toward broad hips.

shoulders are by your hands on the parallel handles. Bend your knees and lower your body to a slow count of ten. You'll feel it more in your back if you lean into the machine with your torso as you descend. Climb back to the top and repeat the negative-only repetitions. Stop the exercise only when you cannot control your lowering.

After a brief period of rest, you'll want to give your best effort to exercises 7–14. Each has been detailed previously. Your primary objective is to do one or two repetitions more than the last time you did the movement in each listed exercise. You can do it, so apply yourself.

SIZE, STRENGTH, AND POWER

This six-exercise back routine added a significant amount of muscle to Eddie's lats. Plus his overall back strength and power increased dramatically. Hit it hard for three consecutive workouts and it will do the same for you.

Right. No one has ever accused Frank Richard of having a narrow back.

Overleaf, left. Shoulder shrugs can be performed with a barbell or dumbbells.
Overleaf, right. Juliette Bergman and Erica Giesen in back-to-back confrontation.

WEEK 7: THICKENING YOUR CHEST

With a touch of good-natured sarcasm in his voice, Eddie asked, "What's the lucky muscle group we'll be working on this week?"

Returning the sarcasm, I replied, "Chest, Eddie, chest. We'll be trying to beef up your pecs so the kids at school will quit mistaking you for a flat-chested girl."

"No one accuses me of looking like a girl," Eddie said, tensing his upper body. "Not with these arms. But I will have to admit, Dr. Darden, my chest seems to be my weakest body part."

I mentioned a new double pre-exhaustion chest cycle I had been working on. To gear Eddie up for the chal-

Above. Tony Pearson and Gary Leonard compare chests. *Opposite.* Both single-joint and multiple-joint exercises are needed for complete chest development.

lenge, I showed him some pictures of massive chests I had recently received from Chris Lund. The photos were taken at a bodybuilding contest in Columbus, Ohio.

"Wow," Eddie reacted. "That's Rich Gaspari on the left and Mike Christian on the right. With chests like that, they must have won the contest."

An excellent deduction. Gaspari and Christian finished in first and second place, respectively.

"You know," Eddie continued, "when I first got interested in working out the old-timer who had the thickest chest was Arnold. Man, were his pecs massive!"

"Arnold Schwarzenegger!" I laughed. "You're calling him an old-timer?"

"Well, he's certainly older than me . . . . and Gaspari and Christian."

With hesitation I led up to my next question.

"I guess you're right, Eddie. But how do you classify me? I'm at least five years older than Arnold."

"You are?" Eddie exclaimed with a surprised expression. "Then you must be over fifty."

I was not sure who should have felt less flattered. But I didn't want to press the issue.

"Never mind. Let's get to your chest routine," I said. "I think you'll find that it will work for the old-timers as well as the newcomers, and anybody in between—including wiseasses!"

CHEST ROUTINE

Your chest routine is composed of two double pre-exhaustion cycles followed by a set of push-ups. In all, you'll be doing seven exercises for your chest and seven for the rest of your body.

1. Bench press to neck, immediately followed by
2. Bent-armed fly, immediately followed by
3. Dip, negative only
4. Bench press, immediately followed by
5. Bent-armed fly, immediately followed by
6. Dip, negative only, immediately followed by
7. Push-up
8. Leg press

GENERAL SERVINGS AND SAMPLE MENU FOR WEEK 7

GENERAL SERVINGS (TOTAL CALORIES: 3,700 PER DAY)

Basic Food Group	Recommended Daily Servings
Meat	6.00
Milk	6.00
Fruit/Vegetable	12.00
Bread/Cereal	12.00
Other Foods	9.00

SAMPLE MENU (TOTAL CALORIES: 3,700 PER DAY)*

Breakfast	Calories:	855
McDonald's		
1 Egg McMuffin		325
1 order of hash browns		125
2 cartons of milk		300
2 packets of jelly (take 2 more packets with you for later use)		105

Lunch	Calories:	1105
Taco Hut		
1 taco salad		140
1 beefy burrito		445
1 enchilada		370
1 carton of milk		150
(take 4 packets of saltine crackers with you for later use)		

Dinner	Calories:	1475
Pizza Hut		
Super Supreme Pizza, thin and crispy 5 slices		1325
1 carton of milk		150

Snack	Calories:	265
8 crackers, saltine (from above)		80
2 packets of jelly (from above)		105
1 apple (from home)		80

* When away from home, you can still get the required servings by carefully selecting foods from fast-food establishments.

WEEK 7: CHEST ROUTINE

	Date:	Date:	Date:
1. Bench press to neck			
2. Bent-armed fly			
3. Dip, negative only			
4. Bench press			
5. Bent-armed fly			
6. Dip, negative only			
7. Push-up			
8. Leg press			
9. Leg curl			
10. Leg extension			
11. Calf raise			
12. Nautilus pullover			
13. Biceps curl			
14. Triceps extension			

Erica Giesen and Bev Francis hit a side chest pose for the judges.

Rich Gaspari and Mike Christian were first- and second-place winners in the 1986 World Professional Championships.

9. Leg curl
10. Leg extension
11. Calf raise
12. Nautilus pullover
13. Biceps curl
14. Triceps extension with one dumbbell held in both hands

● BENCH PRESS TO NECK: This is a terrific exercise for your upper chest. Because you are lowering the bar to your neck, rather than your chest, you'll need about 25 percent less weight than you'd normally handle in the bench press. Position your hands slightly wider than the width of your shoulders and bring the bar over your chest. Lower the barbell slowly, keeping your elbows wide, and lightly touch it to your neck. Press the bar smoothly to the top position. Repeat for maximum repetitions. Move quickly into position for the bent-armed fly.

● BENT-ARMED FLY: Grasp two dumbbells, lie back on the bench, and press them over your chest. Lower the dumbbells slowly by bending your elbows and shoulders. Keep your hands, elbows, and shoulders in line. Stretch in the bottom, and smoothly move the dumbbells back to the straight-armed position over your chest. Repeat. When you can no longer perform a repetition, sit up, place the dumbbells on the floor, and run to the parallel bars for dips.

● DIP, NEGATIVE ONLY: The first two exercises should have thoroughly pre-exhausted your pectoral muscles. The dips will carry the stimulation a step further. Place a chair or sturdy box between the dip bars. Climb to the top position. Straighten your arms, remove your feet from the chair, and stabilize your body. Lower slowly to the stretched position. This negative movement should take eight to ten

Above. Seated dips can be performed on a new leverage machine called the bench press plus.
Left. Bench press: Always lift and lower the barbell slowly.

Opposite, above. Bench press to neck: Lowering the bar to your neck increases the range of movement, which makes the exercise harder.
Opposite, below. Bent-arm fly: Lower the dumbbells slowly to the stretched position.

seconds. Quickly climb back to the top and repeat for the optimum number of repetitions. Take a breather and ready yourself for the next cycle.

• BENCH PRESS: Don't be afraid to go heavy on the bench. Grind out as many repetitions as you can in the normal style. Immediately do another set of the bent-armed fly.

• BENT-ARMED FLY: Do not reduce the weight. Employ the same dumbbells as you used for the first cycle. Make your pecs stretch, contract, and pump with blood as you force them to grow to massive size. After your final repetition, get back to the parallel bars.

• DIP, NEGATIVE ONLY: Your goal is to do the same number of repetitions as you did previously on the dip. You probably won't be able to make it, but you should try. Be sure to stretch fully in the bottom position and climb as fast as possible back to the top. When you've exhausted your negative strength, stop. Lie on the floor and get into the push-up position.

• PUSH-UP: If you've really extended yourself on the other three exercises, then you'll be lucky to get four push-ups. Do as many as you can, regardless of the number. If you can do ten or more, you've probably rested too long between getting from the dip back to the floor. Remember, anything longer than three seconds is cheating.

Your chest should now be thoroughly pumped. That's an indication that something is happening, and that something will happen: growth! Walk around for several minutes until your pump recedes slightly. Then, perform the rest of your workout, exercises 8–14, in the standard manner.

PROPER EXERCISE, PROPER DIET

A thick, massive chest is not easy to develop. But it can be done with proper exercise and proper diet. Practice the double pre-exhaustion cycles and the basic food guidelines in this chapter and you'll be pleasantly rewarded.

Right. Frank Richard demonstrates two versions of flys performed with cable pulleys.

The thick, massive torso of
Ed Kawak.

WEEK 8: BLASTING YOUR ARMS

No one has to convince bodybuilders about the importance of building big arms. The biceps and the triceps have always been, and will always be, the favorite body part of bodybuilders throughout the world.

With the popularity of arm exercises, you'd think that many bodybuilders would be satisfied with the size of their biceps and triceps. They certainly work them enough.

Yet I've never met a single bodybuilder who was completely pleased with his arm development. No sir, not a single one!

In my opinion, your arm dissatisfaction is directly related to two factors.

Above. Bertil Fox has some of the most massive biceps and triceps in the business. *Opposite.* Building massive arms is a product of training less, but training harder. Plus, greater attention must be given to controlling the muscular contraction.

GENERAL SERVINGS AND SAMPLE MENU FOR WEEK 8

GENERAL SERVINGS (TOTAL CALORIES: 3,800 PER DAY)

Basic Food Group	Recommended Daily Servings
Meat	6.00
Milk	6.50
Fruit/Vegetable	13.00
Bread/Cereal	12.00
Other Foods	9.00

SAMPLE MENU (TOTAL CALORIES: 3,800 PER DAY)

Breakfast	Calories:	900
4 ounces of oatmeal, instant		400
4 teaspoons of brown sugar		70
2 slices of French toast, frozen		180
4 teaspoons of syrup		70
2 cups of milk, skim		180

Lunch	Calories:	1115
1 cup of tomato juice		50
Salmon patty: 6 ounces of salmon, drained		280
1 egg		80
2 tablespoons of celery, diced		3
2 tablespoons of onion, diced		2
Combine all above ingredients into a patty and fry using:		
2 teaspoons of butter or margarine		70
2 slices of bread		160
1 cup of beets		60
1 cup of noodles, cooked		160
½ cup of blueberries, fresh or frozen		40
½ cup of yogurt, plain, low-fat		60
1 cup of milk, whole		150

Dinner	Calories:	1350
8 ounces of pork roast, lean		800
1 baked potato		100
1 tablespoon of butter or margarine		105
1 cup of beets		60
½ cup of onions, cooked		20
½ cup of carrots, cooked		25
3 dinner rolls, small		240

Snack	Calories:	435
Chocolate milkshake: 1½ cups of milk, skim		135
1½ cups of ice milk, chocolate		300

WEEK 8: ARM ROUTINE

	Date:	Date:	Date:
1. One repetition chin-up			
2. Biceps curl			
3. One-repetition dip			
4. Triceps extension			
5. Leg extension			
6. Leg curl			
7. Reverse leg raise			
8. Shoulder shrug			
9. Lateral raise			
10. Nautilus pullover			
11. Bench press			
12. Stiff-legged deadlift			
13. Wrist curl			
14. Reverse wrist curl			

First, your biceps and triceps are probably your most overworked muscle groups. You'd get much better results if you trained them less. Second, the vast majority of the repetitions you do for your arms are performed too quickly. Instead of bringing into action the maximum number of muscle fibers during each repetition, you probably move much too fast through the range of movement. Thus you have an ample supply of untapped, dormant muscle fibers that could be stimulated to grow under the right circumstances and with the correct exercises.

The right circumstances and exercises are in abundant supply in this chapter. For accelerated arm growth, all you have to do is follow precisely this tried-and-proved routine.

MASSIVE ARM ROUTINE

I first described this unusual arm routine in *High-Intensity Bodybuilding* in 1984. It is hard, brief, and executed in a slow, deliberate manner. You'll do four exercises for your biceps and triceps and ten exercises for your other body parts.

1. One-repetition chin-up (thirty to sixty seconds raising and thirty to sixty seconds lowering), immediately followed by
2. Biceps curl

3. One-repetition dip (thirty to sixty seconds raising and thirty to sixty seconds lowering), immediately followed by
4. Triceps extension with one dumbbell held in both hands

5. Leg extension
6. Leg curl
7. Reverse leg raise
8. Shoulder shrug
9. Lateral raise with dumbbells
10. Nautilus pullover
11. Bench press
12. Stiff-legged deadlift
13. Wrist curl
14. Reverse wrist curl

ONE-REPETITION CHIN UP: The objective of the one-repetition chin-up is to make a single repetition as intense and as slow as possible. Such a style of training eliminates momentum, and in doing so serves to isolate those hurried-through points in your range of movement. Most body-

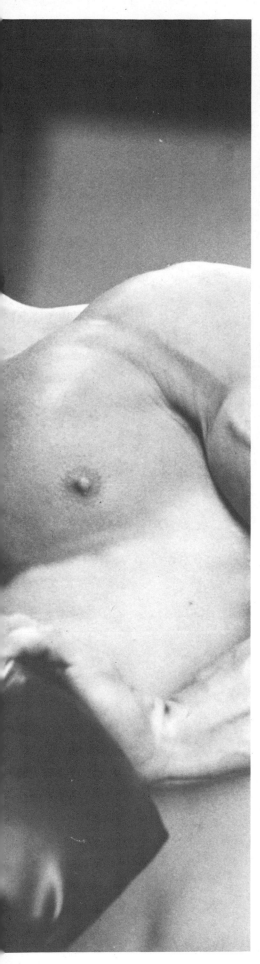

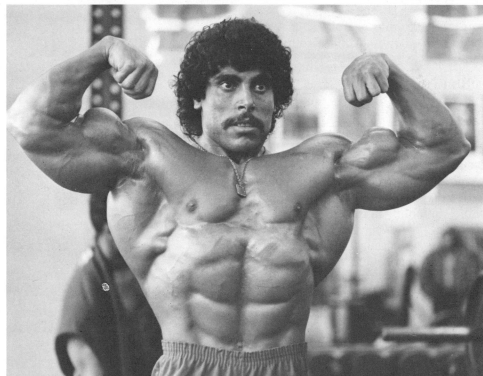

Preceding page, left and above. Greg Comeaux trains his arms with controlled repetitions.

Above. Check the arms and torso of Ed Kawak.

builders perform chin-ups in such a fast manner, using excessive body momentum, that the involved muscles are only partially worked. The slow, one-repetition chin-up allows you to isolate the biceps more thoroughly. Effective biceps isolation leads to better and more complete development.

From a hanging, underhand position with arms straight, take as long as possible to bend your arms and get your chin over the bar. Try to move a fraction of an inch and hold, another fraction of an inch and hold, and so on. Remain in each position briefly (without lowering) and move up inch by inch until your chin is above the bar. Have a friend who has a watch with a second hand call out the time in seconds (five, ten, fifteen, twenty) to you as the exercise progresses.

Once you've achieved the top position, lower yourself in exactly the same manner. Again, a friend or a training partner should call out your time in seconds. Begin this movement with thirty seconds up and thirty seconds down. Add five seconds to both the positive and negative phases each workout. When you can perform sixty seconds down, attach a 25-pound dumbbell around your waist to make the exercise harder. After this unique chin-up, run to the biceps curl.

BICEPS CURL: Doing curls immediately after the one-repetition chin-up will reduce your strength in the barbell curl approximately 50 percent. In other words, you should use about half the resistance you would normally handle. Grasp the barbell with an underhand grip and stand erect. Curl the barbell smoothly in the strictest possible form. Lower slowly to the bottom. Repeat in perfect form for maximum repetitions. Loosen your form and do two more repetitions. Cheat just enough to get past the sticking point. Your biceps should now be under intense pain as you place the barbell on the floor. Have a quick drink of water and get ready to work your triceps.

ONE-REPETITION DIP: The one-repetition dip is performed in a similar fashion to the one-repetition chin-up. Start the dip in the bottom, stretched position. Take thirty to sixty seconds to move to the top and an equal

Above. Bill Pearl, at fifty-six years of age, still has a pair of the most massively developed arms of all time.
Below. Triceps extension: When you feel an intense burn, you'll know growth is being stimulated.

Opposite. Biceps curl: Make your biceps feel the contraction clear to the bone.

amount of time to lower. Your training partner should make sure that he paces you properly by calling out your raising and lowering time in seconds. Next is the triceps extension.

TRICEPS EXTENSION WITH ONE DUMBBELL HELD IN BOTH HANDS: Because of the pre-exhaustion of your triceps, you'll only need about half the resistance that you would normally use. Grasp one dumbbell in both hands, and start performing strict triceps extensions. Keep your upper arms in a vertical position with your elbows stabilized by your ears. Lower and raise the dumbbell behind your head for as many repetitions as possible. When you can no longer do the repetitions strictly, cheat two more up and concentrate on the lowering. Try to squeeze out every bit of negative strength on your last lowering repetition.

Take a several-minute break. As you are resting, your arms should be pumped to a degree that you've seldom achieved in the past. Shake them out and prepare yourself to proceed with the workout.

Give exercises 5–14 your best effort. Try to progress by at least one repetition on each exercise. And remember, your arms grow best not by segmenting your body with various split routines, but by working your entire body at each training session.

ADD ½ INCH TO YOUR ARMS IN A WEEK

Blast your arms with this routine for a week, and your biceps and triceps will *grow!* Eddie added exactly ½ inch of solid muscle to each arm. You can expect the same, or perhaps better, results.

Right. **The mass and muscularity of Wilf Sylvester's body are evident in this picture.**

Opposite. **Work your forearms at the end of your routine by doing wrist curls and reverse wrist curls.**

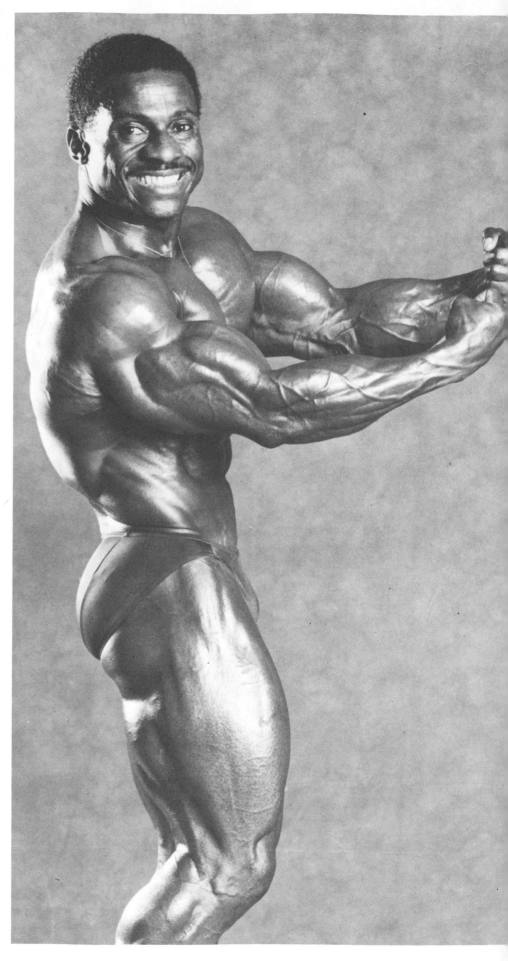

9

WEEK 9: DEFINING YOUR MIDSECTION

Abdominal muscularity is primarily related to the amount of fat that you have around your waistline. Most of this fat is connected to your skin, and its thickness can be measured with a skinfold caliper.

One of the major objectives of this ten-week program is for you to get bigger—by building your muscles, not by getting fatter. That's why it is important to monitor your skinfold thicknesses or keep records of your relaxed and contracted arm measurements (see page 23). If you are gradually getting fatter, as indicated by the caliper or arm measurements, then you must reduce the number of calories you are eating each day.

If you are fairly lean, however, and do not have excessive layers of fat around your waist, the rest of this chapter can show you how to strengthen, build, deepen, and define your abdominal muscles.

Compare the midsection poses of Rich Gaspari and Ed Kawak.

GENERAL SERVINGS AND SAMPLE MENU FOR WEEK 9

GENERAL SERVINGS (TOTAL CALORIES: 3,900 PER DAY)

Basic Food Group	Recommended Daily Servings
Meat	6.00
Milk	6.00
Fruit/Vegetable	13.00
Bread/Cereal	13.00
Other Foods	9.50

SAMPLE MENU (TOTAL CALORIES: 3,900 PER DAY)*

Breakfast	Calories:	840
Instant breakfast, any flavor, 8 ounces of milk, whole		290
Frozen breakfast, such as Swanson's scrambled eggs, sausage, & coffee cake		420
1 cup of orange sections, fresh or canned		130

Lunch	Calories:	1290
Ravioli, canned, beef in meat sauce: 2 cups		500
2 tablespoons of Parmesan cheese		70
1 cup of cranberry/apple juice		145
3 slices of bread		240
3 teaspoons of butter or margarine		105
1 cup of peas, sweet		140
1 cup of milk, skim		90

Dinner	Calories:	1170
8 ounces of breaded frozen fish cakes		600
1 cup of applesauce, unsweetened, sprinked with cinnamon		100
1 cup of grapes, seedless		80
2 dinner rolls, small		160
1½ teaspoons of butter or margarine		50
2 cups of milk, skim		180

Snack	Calories:	600
Root-beer float: 1 12-ounce bottle of root beer		150
1½ cups of ice milk, vanilla		300
3 cups of popcorn, plain		150

If you are too busy or too tired to cook, you can still get the required servings from using canned or frozen foods.

WEEK 9: MIDSECTION ROUTINE

	Date:	Date:	Date:
1. Hanging leg press			
2. Trunk curl			
3. Side bend			
4. Stiff-legged deadlift			
5. Nautilus pullover			
6. Chin-up, negative only			
7. Leg extension			
8. Leg curl			
9. Leg press			
10. Calf raise			
11. Lateral raise			
12. Dip			
13. Pulldown behind neck			
14. 4-way neck			

Extreme muscularity, like that of Frank Richard, is due primarily to a low level of subcutaneous fat.

The deeply etched abdominals of Ed Kawak.

ABDOMINAL ROUTINE

Your midsection workout consists of three pre-exhaustion cycles. After that you do eight other exercises for your legs and upper body. Here's the entire routine.

1. Hanging leg raise, immediately followed by
2. Trunk curl

3. Side bend, immediately followed by
4. Stiff-legged deadlift

5. Nautilus pullover, immediately followed by
6. Chin-up, negative only

7. Leg extension
8. Leg curl
9. Leg press
10. Calf raise
11. Lateral raise
12. Dip
13. Pulldown behind neck
14. 4-way neck

● HANGING LEG RAISE: Hang from an overhead bar. Your hands should be shoulder-width apart. Raise your legs smoothly and touch your toes to the bar. Lower your legs slowly to the bottom position. Do not bend your arms during the raising or lowering. Keep them straight throughout the movement. Repeat until you can no longer touch your feet to the bar. Drop to the floor and lie on your back for the trunk curl.

● TRUNK CURL: With your heels up close to your buttocks and your knees wide, perform trunk curls for as many repetitions as possible. Done strictly and deliberately, you should feel a concentrated burn in your midsection.

● SIDE BEND: With a heavy dumbbell in your right hand, bend laterally to your right side. Return to the erect position and repeat for maximum repetitions. Switch the dumbbell to your left hand and repeat the exercise to your left. Then do the stiff-legged deadlift.

● STIFF-LEGGED DEADLIFT: Doing deadlifts immediately after side bends will force your lower back muscles to a deeper level of fatigue. Do as many repetitions as you can in good form. Rest for several minutes before doing the pullover.

● NAUTILUS PULLOVER: The pull-

Juliette Bergman and Erica Giesen pose their abdominals at the 1986 Ms. International contest. Erica won the contest and Juliette placed second.
A sit-up, as shown above, involves the iliopsoas muscles of the hips as much as it does the abdominals. To eliminate the iliopsoas muscles from the exercise you must *not* anchor your feet or knees. Your feet should be near your buttocks and your knees apart. Doing so limits your range of movement and focuses the work on your abdominals.

over strongly activates your abdominals, especially if you lean forward slightly in the contracted position. Perform as many repetitions as possible; then loosen your form and do two more repetitions. Unbelt yourself from the machine and move quickly to the multi-exercise machine for negative chins.

● CHIN-UP, NEGATIVE ONLY: You'll be surprised how much you feel negative chins in your abdominals when you do them last in this series of six exercises. With your chin over the crossbar, take eight to ten seconds to lower your body to the stretched position. Do not be afraid to use a heavy resistance, if you can, on this exercise. Repeat until you can no longer control the downward movement.

Rest for two or three minutes. Then do the remaining eight exercises for your legs and upper body.

HITTING ALL ANGLES

Do not perform this six-exercise, concentrated routine for your midsection for more than three consecutive workouts. It's easy to overwork your abdominals. Overworking your abdominals will limit your progress throughout your entire body.

Furthermore, spot reduction of waistline fat by exercising your abdominals is impossible. Get it out of your mind that you have to do sets of high-repetition, low-intensity abdominal exercises to have a muscular midsection. That is simply not true.

Strengthen your midsection from all angles: front, sides, and back. Get rid of your excess body fat by limiting your dietary calorie intake. And soon you'll have a well-defined waistline of which you can be proud.

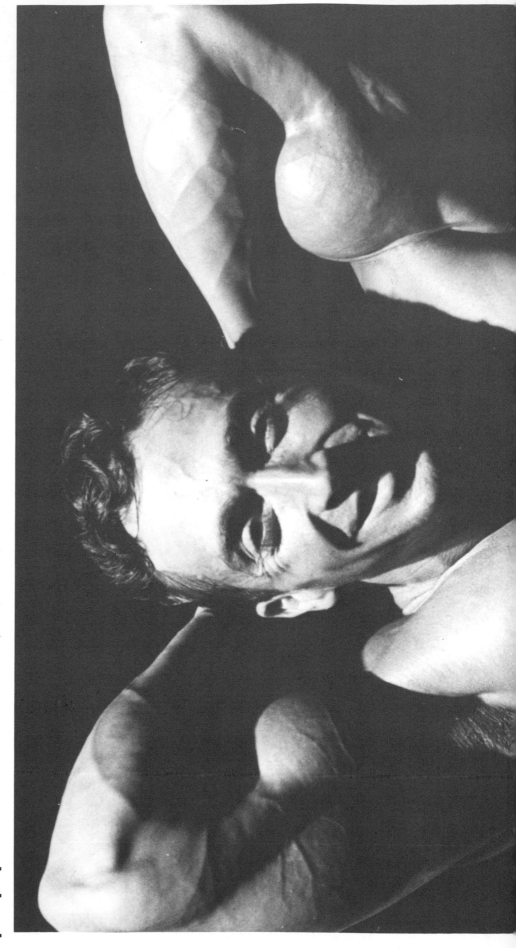

Mark Bolduc tenses his midsection.

Overleaf. Gary Leonard is shown doing an abdominal vacuum.

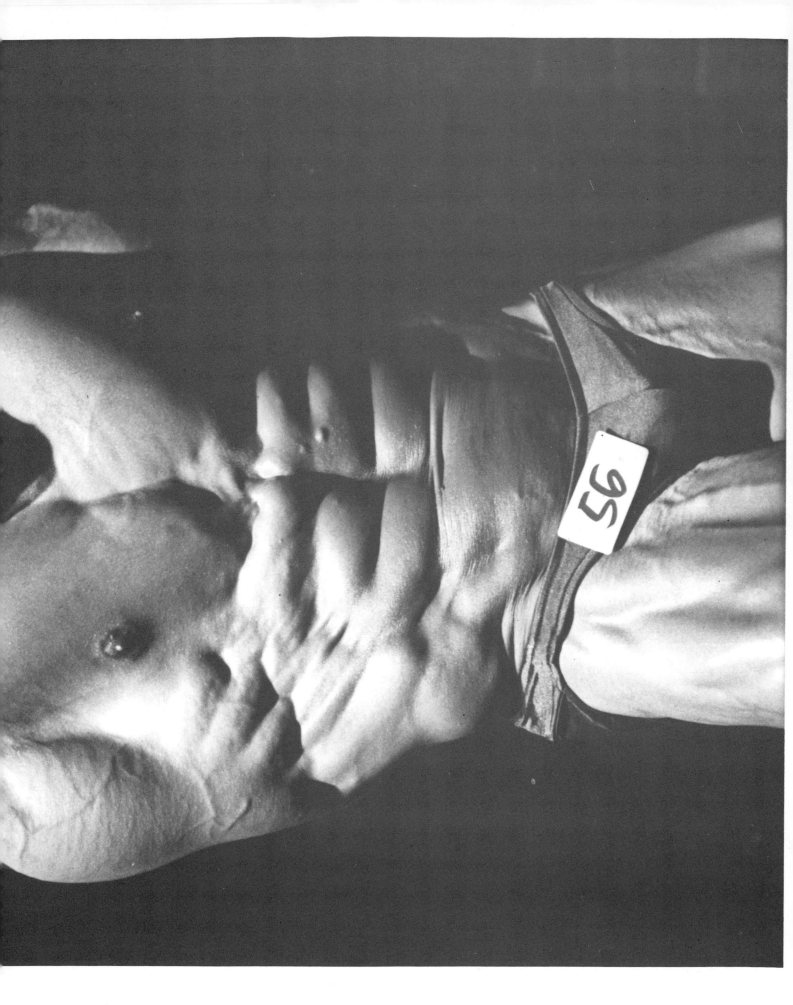

Getting as strong as possible on a few basic exercises is a key factor in developing massive muscles in the most efficient manner.

GENERAL SERVINGS AND SAMPLE MENU FOR WEEK 10

GENERAL SERVINGS (TOTAL CALORIES: 4,000 PER DAY)

Basic Food Group	Recommended Daily Servings
Meat	6.50
Milk	6.50
Fruit/Vegetable	13.00
Bread/Cereal	13.00
Other Foods	9.75

SAMPLE MENU (TOTAL CALORIES: 4,000 PER DAY)

Breakfast	Calories: 1200
4 slices of bread, toasted	320
4 tablespoons of peanut butter	360
2½ tablespoons of jelly or preserves	140
½ cup of prunes, with pits, unsweetened	135
1½ cups of milk, skim	135

Lunch	Calories: 1030
Egg salad: 2 eggs, boiled	160
1 tablespoon of mayonnaise	100
Combine ingredients and place on: 3 leaves of lettuce	10
Macaroni and cheese: 1 cup of noodles, cooked	190
1½ ounces of cheese	150
2 tomatoes	70
1 cup spinach, cooked	40
1 cup of chow mein noodles	220
1 cup of milk, skim	90

Dinner	Calories: 1280
7½ ounces of steak, lean, broiled	500
1 cup of mushrooms	20
2 pieces of corn on the cob	140
4 slices of garlic bread using Italian or French-style bread and	
2¾ teaspoons butter or margarine and garlic salt	385
10 french fries, 2-inches-long	155
1 cup of fruit cocktail, water packed	80

Snack	Calories: 490
Blueberry milkshake: 2 cups of milk, skim	180
1 cup of ice milk, vanilla	200
½ cup of blueberries, fresh or frozen	40
20 pretzels, small stick type	70

WEEK 10: BASIC ROUTINE

	Date:	Date:	Date:
1. Leg extension			
2. Leg curl			
3. Full squat or leg press			
4. Reverse leg raise			
5. Calf raise			
6. Lateral raise			
7. Overhead press			
8. Nautilus pullover			
9. Pulldown behind neck			
10. Bench press			
11. Bent-over row			
12. Triceps extension			
13. Biceps curl			
14. Stiff-legged deadlift			

I was assessing Eddie's nine weeks of labor. I had to ask him about an important area—his body weight.

"Almost one seventy-five," was Eddie's curt reply.

"What do you mean, *almost one seventy-five*?" I questioned. "Do you weigh one seventy-four and a half, one seventy-four, one seventy-one, or what? And is that with your clothes on or off?

"Okay," he answered with a calculated grin on his face. "I weigh exactly one seventy-four and a quarter pounds, stripped naked, with an empty stomach. Plus, I took a leak before I stepped onto the scale."

"Great," I said. "That's exactly what I wanted to know."

I then launched into a sermon.

"Being specific seems to be a lost art among teenagers today. Ask a question and you get *ahs, wells,* and *you knows* until you can hardly stand it."

I then realized that this phenomenon is not relegated only to teenagers.

"The same thing is true when you try to communicate with most bodybuilders concerning how much they can bench press, or the size of their arms. Talk about double talk and evasive and exaggerated numbers!"

Sermon over, but I hoped the message was clear.

"Let's get specific for Week Ten," I said. "Eddie, we're going back to the basic, mass-building routine that we used for the first two weeks. Doing so will allow us to compare your strength now with what it was then."

Eddie wanted to know if the routine would add muscle to his body, so that he could surpass the 175-pound mark.

I assured him that it would.

"Just follow the routine exactly as planned, and eat all of the recommended servings. That should do the trick. But as an insurance policy, before you step on the scales next week—don't take a leak!"

Mike Christian has what it takes: size, shape, muscularity, and posing ability.

Many people believe Juliette Bergman possesses the best genetic potential of any woman bodybuilder competing now.

Preceding page, left and above. **Always emphasize the basic exercises, such as the leg extension, squat, pullover, and biceps curl.**

BASIC MASS-BUILDING ROUTINE

Here's your basic routine for Week 10:
1. Leg extension
2. Leg curl
3. Full squat or leg press
4. Reverse leg raise
5. Calf raise
6. Lateral raise with dumbbells
7. Overhead press
8. Nautilus pullover
9. Pulldown behind neck
10. Bench press
11. Bent-over row
12. Triceps extension with one dumbbell held in both hands
13. Biceps curl
14. Stiff-legged deadlift

None of the exercises, or groups of exercises, are performed in a pre-exhaustion manner. So it is not necessary to rush, in less than three seconds, from one exercise to the next. But at the same time, there is no need to rest longer than twenty to thirty seconds between most exercises. After nine weeks of training, you should be in good enough condition to work through all fourteen exercises in thirty minutes or less.

THE TEN-WEEK CHALLENGE

How much bigger, stronger, and leaner are you now that you've completed the ten-week program? In the next chapter, you'll find out how well you have met the ten-week challenge. Also, you'll get a chance to compare your results with Eddie's.

Mike Quinn and Dave Hawk understand the importance of hard, brief exercise.

EVALUATING YOUR
RESULTS

Above. After 10 weeks your forearms should be significantly larger and stronger. *Opposite.* If you've adhered to the 10-week program exactly as directed, you should have added at least 4 inches to your chest and back, and 2 inches on your upper arms.

For this chapter to be beneficial to you, you'll need to get your after testing completed as soon as possible. If you performed your last workout for Week 10 on Friday, then your final measurements, weigh-in, strength tests, and photographs should be done on the following Monday or Tuesday.

Let's review what you need to do.

CIRCUMFERENCE OF BODY PARTS

Go back and read carefully on pages 21–23 the directions for taking accurate circumference measurements. Be sure the same tape is used for both the before and after measurements. Take duplicate readings at each of the described sites and use the average figure as your circumference score. Subtract the before score from the after score and make a list of these differences.

How do your increases compare with Eddie's? You'll be able to tell by examining the chart below.

Eddie's Measurements

Body Site	Before	After	Inches Gained
Neck	15⅝	16¼	⅝
Right upper arm	14⅞	16	1⅛
Left upper arm	14½	15⅜	⅞
Right forearm	12½	13	½
Left forearm	12⅛	12¾	⅝
Chest	41	45	4
Waist	31⅜	31⅜	
Hips	36⅜	38¼	1⅞
Right thigh	21¾	24	2¼
Left thigh	21¾	24	2¼
Right calf	14¾	15⅛	⅜
Left calf	14⅞	15½	⅝
Total Inches Gained			**15⅛**

It is interesting to note that Eddie gained the most inches in his chest and thighs. He added 4 inches to his chest and 2¼ inches to each thigh. He also added 1⅛ inches to his right upper arm and ⅞ of an inch to his left upper arm. Equally important was the fact that his waist measurement did not increase. It stayed the same, which is an indication that he did not get fatter. Eddie's total inches gained, derived by summing the twelve scores, was 15⅛ inches.

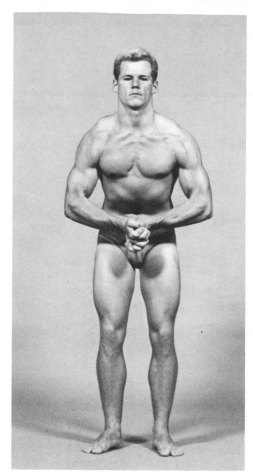
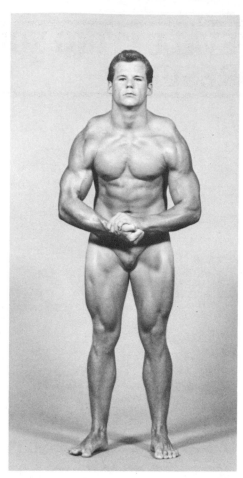
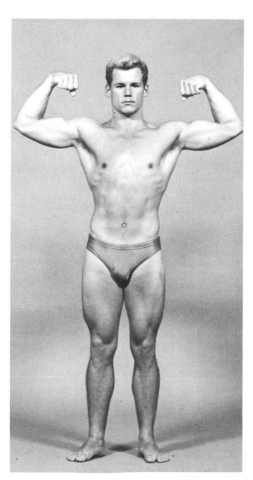
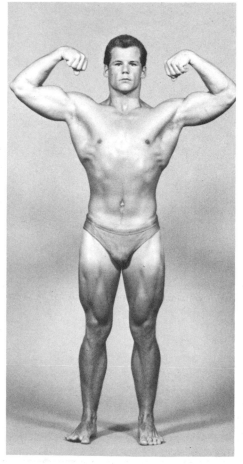

BODY WEIGHT

You should already be aware of body weight gains from your periodic weigh-ins before your workouts. Make certain your after weigh-in is done under the same conditions as your before weigh-in. For example, you should use the same scale, wear the identical clothes (or go nude), and weigh at approximately the same time of day.

For comparison purposes, Eddie, at a height of 5 feet 8 inches, weighed 160 pounds at the start of the program. At the conclusion of ten weeks, he weighed 176 pounds. In ten weeks he gained 16 pounds, or an average of 1.6 pounds per week.

SKINFOLD TESTING

If you have access to a skinfold caliper, be sure to repeat your measurements at exactly the same location as before. If you use a Lange caliper, you'll want to compare your findings to Eddie's.

Eddie's Skinfold Values

	BEFORE (in millimeters)	AFTER (in millimeters)
Chest	9	7
Waist	14	11
Thigh	8	6
Total	31	24
Percentage of Body Fat	8	6

Eddie reduced the sum of his three skinfold readings from 31 millimeters to 24 millimeters. Using a special nomogram, these figures convert to 8 percent and 6 percent body fat. He lost 2 percent of his fat over the ten-week program. This translates to a fat loss of 2.2 pounds.

Thus, if Eddie gained 16 pounds of body weight and lost 2.2 pounds of fat, he actually added 18.2 pounds of muscle to his body.

If you do not have a skinfold caliper handy, you'll have to make do with comparing the differences between your relaxed and contracted upper arm measurements. No conversion tables

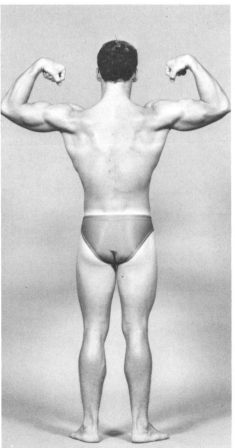
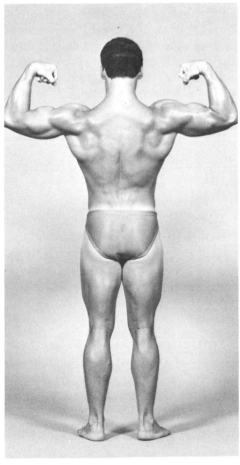

Before-and-after comparisons of Eddie Mueller.

are available for this test. You'll simply have to make sure that the differences between your relaxed and contracted upper arm measurements, compared over ten weeks, are getting larger. A larger measurement means you are getting leaner. It means you are less fat now than you were previously.

For comparison purposes, Eddie's differences increased by ⅛ of an inch in ten weeks. His before difference was 1½ inches and his after difference was 1⅝ inches.

STRENGTH TESTS

Since you tested yourself ten weeks ago to determine the best repetition scheme to use throughout the ten-week program, all you have to do is retest yourself with the following modified guidelines.

1. Use only three exercises for your lower body and three exercises for your upper body. I'd recommend the leg extension, leg curl, leg press, Nautilus pullover, bench press, and biceps curl.
2. Determine your one-repetition maximum on the first exercise.
3. Rest at least five minutes.
4. Determine your one-repetition maximum on the other five exercises using the same method.
5. Compare your before strength with your after strength.

The best way to compare your before strength with your after strength is to find your average strength score for both your lower and upper body. Your beginning and ending lower body strength score is determined by adding together your one-repetition maximums on the leg extension, leg curl, and leg press and then dividing by three. This is your average lower body strength when you started and ended the program.

Adding together your one-repetition maximums for the upper body exercises—Nautilus pullover, bench press, and biceps curl—and then dividing by three will provide you with your average upper body strength at the beginning and the end.

To calculate your percentage increase in lower body strength, subtract the before average from your after average. That amount divided by the

before average is your percentage of increase. Do the same for your upper body strength. Now compare your strength increases with Eddie's.

Eddie's Strength Tests (in pounds)

	Before	After	% Increase
Leg extension	170	270	59
Leg curl	155	220	42
Leg press	230	380	65
Average lower body strength	185	290	57
Nautilus pullover	150	195	30
Bench press	200	310	55
Biceps curl (leverage machine)	85	125	47
Average upper body strength	145	210	45

You'll notice that Eddie's lower body strength improved an average of 57 percent per exercise. His upper body strength increased an average of 45 percent per exercise. Averaged together, he had approximately a 5 percent increase in strength on each exercise each week for ten weeks.

Since there is a direct relationship between the strength of a muscle and the size of a muscle, it is evident that Eddie's 5-percent strength improvement each week produced 1.82 pounds of muscular growth each week. Thus, in ten weeks, with a 50 percent average improvement in his overall strength, he added 18.2 pounds of muscle to his body.

FULL-BODY PHOTOGRAPHS

Full-body photographs are one of the best ways to evaluate the improvements in your physique over the last ten weeks. The taking of your after pictures must be standardized according to the instructions on page 24.

Although no valid rating scale is available for making before-and-after comparisons with the photographs, your eyes shouldn't require any instructions on how to evaluate the muscle mass improvements—which should be quite obvious.

In comparing the pictures it helps to

use photocopies of the photographs. With photocopies you feel free to measure, mark, and encircle certain areas with different colored pens. Such notes can be a valuable addition to your progress charts.

THE NEXT STEP

How do your after measurements stack up against Eddie's results? Have you reached some of your bodybuilding goals during the last ten weeks? Would you like to go further in your quest for more massive muscles?

The next chapter will show how to continue with your program and eventually reach your genetic potential.

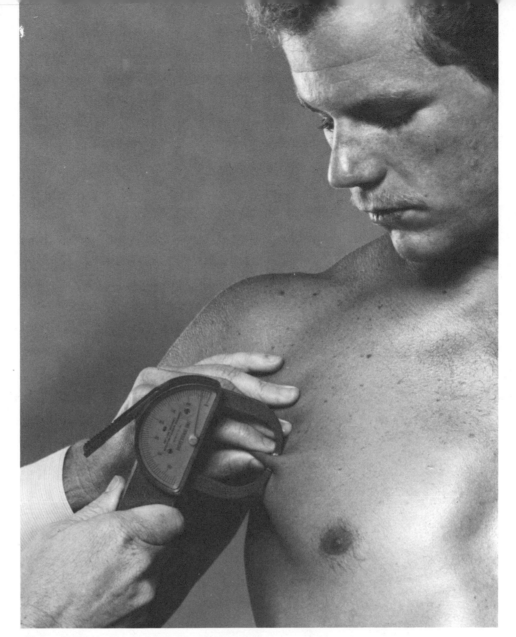

Above. Eddie's total skinfold thickness decreased from 31 millimeters to 24 millimeters. Thus his body fat was reduced from 8 to 6 percent.
Below. In 10 weeks, Eddie improved his strength in the leverage biceps curl by 47 percent.

MAKING MORE PROGRESS

"You deserve a break today," notes the familiar television commercial.

Anyone who has been through a grueling ten-week bodybuilding program deserves a break — or a *brief layoff from training.*

How long should your layoff last? It should last at least a full week. Ten days would be better yet, since training could be terminated on a Friday and resumed on Monday of the second following week. Two weekends of rest can do wonders for your motivation and future progress.

Repeat the Ten-Week Program

After a brief layoff, I put Eddie back on the same ten-week plan. In fact, it is my intention to repeat the ten-week program and layoff at least twice, or until he reaches a significant strength plateau in the majority of the basic exercises.

I believe Eddie has the potential to weigh 200 pounds without adding fat to his existing physique. Repeating the exercise routine once again in the same order is a big step in the direction of reaching that goal.

I also increased Eddie's dietary calories, not by 100 calories per week as before, but by 50 per week. His new calorie recommendations for Weeks 11–20, in outline form, looked like this:

Week 11: 4,050 calories
Week 12: 4,100 calories
Week 13: 4,150 calories
Week 14: 4,200 calories
Week 15: 4,250 calories
Week 16: 4,300 calories
Week 17: 4,350 calories
Week 18: 4,400 calories
Week 19: 4,450 calories
Week 20: 4,500 calories

As with the ten-week program, I would monitor his skinfold thicknesses periodically to make certain the added calories were not making him fatter.

I suggest that you continue your workouts in a similar manner as Eddie. You may want to make minor adjustments according to your strengths and weaknesses. If your chest seems to be lagging behind, you can devote two weeks, instead of only one, to working it. The two weeks, however, should be nonconsecutive weeks. The same situation also applies to your calves, shoulders, or any other body part. Of course, you'll have to remove one of your strongest body-part routines to make room for the lagging muscle group that needs more emphasis.

Super High Intensity

After you have completed the ten-week program and layoff two or three times, you should be strong and advanced enough to progress to a harder, more varied plan. I recommend that you follow the routines in my book, *Super High-Intensity Bodybuilding,* which may be ordered by sending $13.50 to Darden Research, P.O. Box 1016, Lake Helen, FL 32744.

With this book, you'll be well on your way to building the ultimate in massive muscles.

Right and Overleaf. **Both Juliette Bergman and Erica Giesen recognize the importance of keeping their dietary calories in balance with their activity calories.**

Mike Quinn's massive back is one of his most outstanding body parts.

Ed Kawak has no visible weaknesses in his championship physique.

Above. Albert Beckles has the highest-peaked biceps of any professional bodybuilder. *Opposite.* Massive muscular size and strength are produced by hard work, adequate rest, and proper nutrition.

Q. Where can I purchase a skin-fold caliper?

A. Skinfold calipers have been used by physicians, physiologists, and nutritionists for many years to estimate a person's percentage of body fat. A caliper is practical for a physician or clinician to use, but rather expensive for the average bodybuilder, since a good metal pair costs at least $150. Recently, however, inexpensive plastic models — selling for $9.95 or less — have proved to be reasonably accurate when compared to the metal versions.

You may buy various types of skinfold calipers from a local medical-supply house. Or you may purchase them through the mail by writing to Fitness Finders, P.O. Box 507, Spring Arbor, MI 49283.

Q. I take protein supplements to make certain my body is supplied with enough muscle-building materials. Is there a protein supplement that you recommend?

A. Sorry to disappoint you, but protein supplements are not necessary for building massive muscles. This confusion frequently arises because bodybuilding magazines constantly promote the idea that "muscles are made of protein." Scientific investigation reveals that more than 70 percent of a muscle is water and only 22 percent is protein.

Proper exercise, not protein supplementation, is the key factor in muscular growth. If you stimulate your muscles to grow with proper exercise, such as the ten-week routines in this book, then those muscles will grow with no additional protein added to your diet. The reason is that it takes very little protein to produce muscular growth, particularly since so little of it takes place within a given twenty-four-hour period.

For example, there are approximately 100 grams of protein in one pound of muscle. Most bodybuilders would be thrilled to add one to two pounds of muscle to their bodies per week. (Eddie added 1.8 pounds of muscle to his body each week.) A trainee who stimulates two pounds of muscular growth each week would need to have approximately 30 grams of protein per day added to his minimum requirement. The minimum level for most bodybuilders would be between 30 and 40 grams per day.

Thus any hardworking bodybuilder could develop his muscles to maximum size on a daily diet that included from 60 to 70 grams of protein. And surveys show that the average bodybuilder in the United States consumes well over 200 grams of protein per day from ordinary foods, which doesn't include protein supplements. If you're dissatisfied with your bodybuilding results, it is almost certain that the problem is *not* due to lack of protein.

Q. But I keep reading in the bodybuilding magazines that all of the champions eat over 300 grams of protein per day. If such consumption is of no value, why do they keep doing it?

A. You read this in bodybuilding magazines because most of the publishers of the magazines also sell protein supplements. Naturally, the magazines do not want to promote the fact that protein supplements are totally unnecessary in muscle building. The bodybuilders who do not use protein supplements are often paid handsomely to say they do. "Money," the old saying goes, "makes people do and say almost anything."

One of the greatest bodybuilders of all time, Sergio Oliva, frequently consumes pizza and cola drinks after his training sessions. Yet pizza and colas are often condemned in muscle magazines and health food publications as "junk foods."

Nutritional research, however, has proved that the primary nutrients needed by Sergio Oliva to maintain and further develop his muscle mass are calories and water. Pizza and colas would fulfill his needs quite well.

Q. Are you saying that colas and other soft drinks are not junk foods?

A. Food scientists have found that there are no true junk foods. Any food can be labeled as junk if it is excessively consumed. That same food, if eaten in moderation with a variety of other foods, can contribute significantly to nutritional well-being.

Colas and other regular soft drinks are not important sources of nutrients, other than calories and water, nor are they harmful if ingested in moderation. They provide up to sixteen calories per fluid ounce and come with or without caffeine. For the weight-conscious, diet soft drinks provide enjoyment without calories. Soft drinks should be evaluated for what they are: a source of water and liquid refreshment, a thirst quencher, and an accompaniment to food.

Contrary to their advertised image, soft drinks will not make you popular. But enjoying soft drinks is not something to fear or feel guilty about, either.

Q. What about taking free-form amino acids?

A. Amino acids are the building blocks of proteins. There are twenty-two amino acids that are important in human nutrition. Some of these amino acids can be separated into very pure forms and sold as free-form amino acids.

There is no advantage, however, to taking free-form amino acids. They will not help you build muscle faster.

In fact, according to Dr. James Kenney, who has written an excellent report on the subject (*City Sports:* May, 1985) "free-form amino acids are anything but free." For example, 100 grams (about 3.5 ounces) of a popular amino acid powder retails for $26.98. That works out to $122.49 per pound. Dr. Kenney's research shows that you can get the same approximate amount of free-form amino acids by eating ten ounces of chicken breast. When you consume chicken breast, or other protein foods, your pancreas, stomach, and small intestine produce a small army of digestive enzymes that systematically break down the proteins to individual amino acids.

In other words, your body can make more than adequate amounts of free-form amino acids from the protein foods that you eat at each meal. You do not have to pay to have a manufacturer do this for you. Your body can do a much better job of processing, separating, and mixing amino acids than can any supplement company.

Q. Then you're saying that all the advertising that the free-form aminos arginine and ornithine produce "faster-than-ever-before muscle growth" is a bunch of hooey?

A. Yes, that's exactly what I'm saying. Furthermore, the Federal Trade Commission recently entered into a consent agreement with one large manufacturer regarding its advertising for amino acid supplements. The company can no longer advertise that these free-form amino acid products cause greater or faster muscular development because there is no reliable scientific proof that they do. The manufacturer also must offer refunds to people who purchased these products as a result of his misleading advertisements.

It should be pointed out, once again, that a deficiency of protein, or any of the amino acids that make up protein foods, is never seen in bodybuilders who eat anything close to a balanced diet. Regardless of what you read in the muscle magazines, do not waste your money on expensive free-form amino acids or protein supplements. They are totally unnecessary for building massive muscles.

Q. Is there any value in taking raw glandular supplements to promote muscle building?

A. Raw glandular supplements are another of the latest fads among bodybuilders. The claim is that these substances, which are derived from animal tissues or glands, will supercharge your body with additional muscle-building hormones. For example, tablets made from raw beef hearts will give you better circulation, raw pituitary will improve performance, and powdered bull testicles will strengthen your muscles. The substances are advertised by one company as being derived from "specially selected Argentine bovine products."

What is not stated on these labels is the fact that when such substances are taken by mouth, they are digested and destroyed by the stomach and intestines. The claims behind raw glandular supplements are sheer nonsense.

Q. Did Eddie Mueller take steroids during your ten-week program?

A. No, Eddie did not take any of the androgenic-anabolic steroids during the program. He built his body with hard work and nutritious food.

Q. Can I substitute another vegetable for broccoli in the dinner menu for Week 3?

A. Yes, that is fine. In fact, the specific dietary guidelines in Weeks 1–10 are simply examples of meals that meet the general recommendations. You may substitute freely as long as the foods are from the same food group and the calories per serving are similar.

Q. Will the routines in this book work equally well for women bodybuilders?

A. Women have the same number of skeletal muscles as men. Muscular function, regardless of sex, is identical. Male and female muscle requirements for growth are the same. It is important to understand, however, that women do not have the same genetic potential for building muscular size and strength as men. The average man can increase his muscular strength by 300 percent before he reaches his potential. The average woman increases her muscular strength by approximately 200 percent before she reaches hers.

Q. What is the difference between doing a leg press and a squat on the muscles of the lower body? Is one better than the other?

A. In both the leg press and the squat you have movement around three major joints: ankle, knee, and hip. In the squat, however, you have more potential movement around the hip than you do in the leg press. Contact between your thighs and chest in the back position limits your range in the leg press. The squat therefore involves your buttocks and lower back muscles more than the leg press.

Both the squat and the leg press are excellent exercises for your lower body, but if you had to choose between the two, the squat would be the better selection.

Q. You don't seem to place much emphasis on using dumbbells in your routines. Why?

A. Dumbbells are more of a hassle to assemble and disassemble than barbells. But other than that, I've got nothing against them. And they can certainly add variety to your workouts. If you substitute an occasional dumbbell exercise for a barbell one, make sure that you note it on your workout card.

1-10 AGAIN

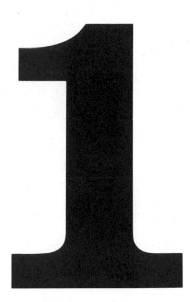

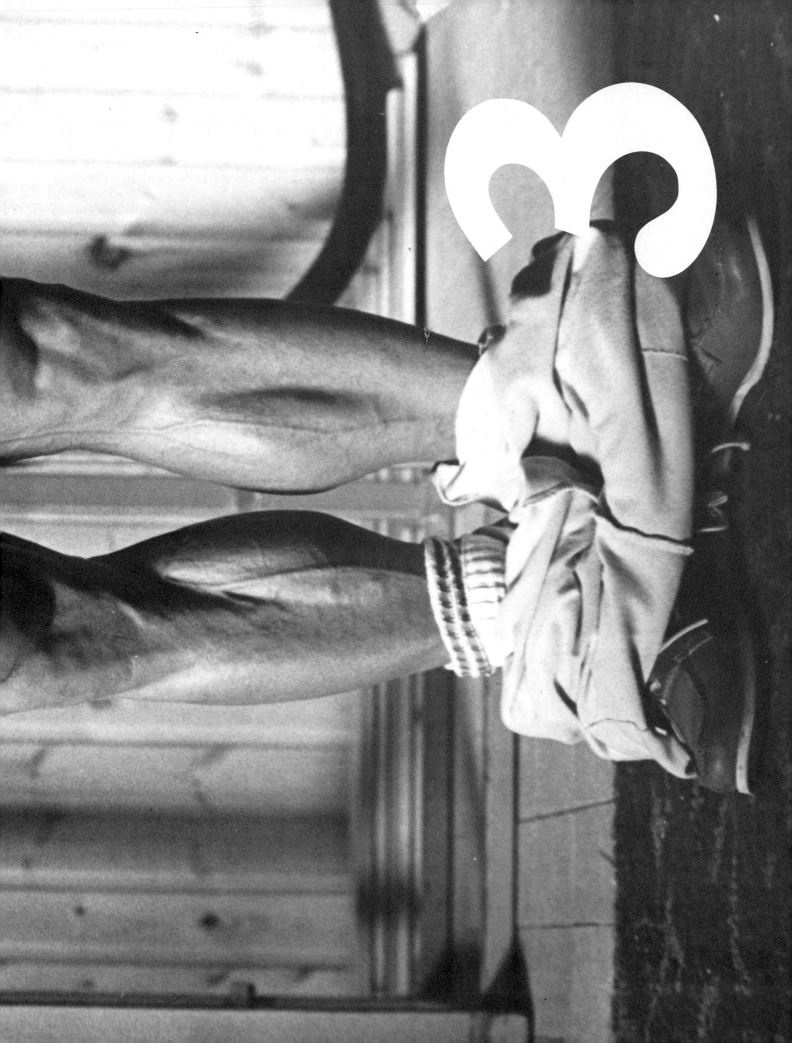

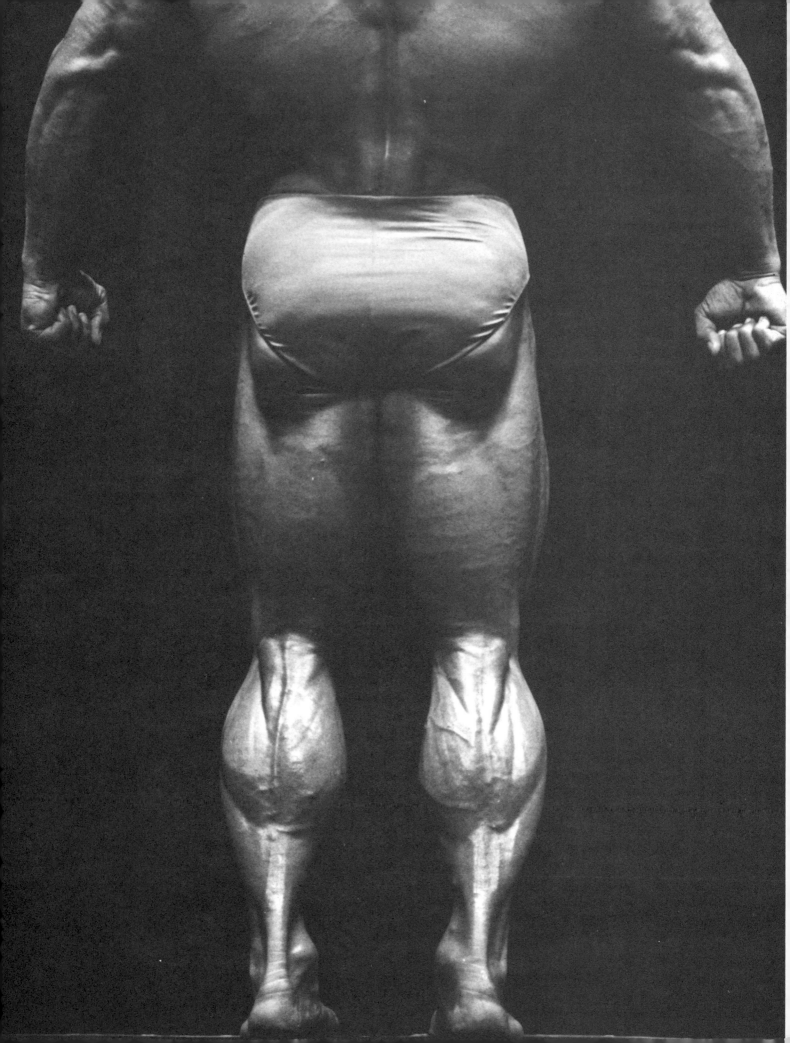

5

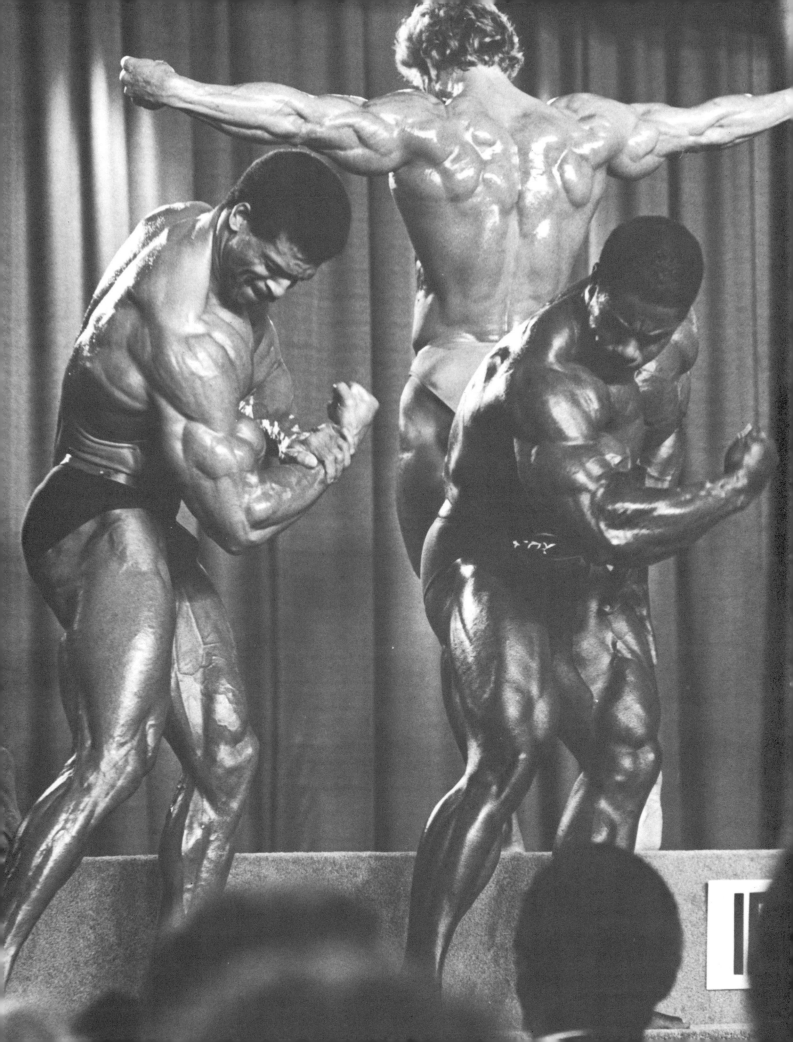

10

CONCLUSION

Building massive muscles requires the application of a careful plan. The plan combines hard, brief, progressive exercise with carbohydrate-rich meals.

This book contains the plan. All you have to do is apply it to your life-style for ten weeks.

Do it now!